Toward an Understanding of the Progenitors of Gamma-Ray Bursts

by
Joshua Simon Bloom

ISBN: 1-58112-169-5

DISSERTATION.COM

Parkland, FL • USA • 2002

Toward an Understanding of the Progenitors of Gamma-Ray Bursts

Copyright © 2002 Joshua Simon Bloom
All rights reserved.

Dissertation.com
USA • 2002

ISBN: 1-58112-169-5
www.Dissertation.com/library/1121695a.htm

Toward an Understanding of the Progenitors of Gamma-Ray Bursts

Thesis by
Joshua Simon Bloom

In Partial Fulfillment of the Requirements
for the Degree of
Doctor of Philosophy

California Institute of Technology
Pasadena, California

2002

(Defended 1 April 2002)

© 2002
Joshua Simon Bloom
All rights Reserved

Acknowledgements

I first encountered my future thesis adviser, Professor Shri Kulkarni, at his colloquium on soft gamma-ray repeaters in 1993. I was a sophomore at Harvard and had just decided to concentrate in astronomy and physics for my undergraduate degree. I remember that afternoon vividly: his talk was explosive and Shri was ebullient and animated. In that one hour, he managed, with great gusto, to capture the essence of the field and left me intensely excited about a subject that I had known little about. I know now that such a talent is rare.

Shri is indeed a rarity. He is at once enthusiastic, brusque, contradictory, logical, forward-thinking, generous, unfathomably demanding, focused, multi-plexed, socio-politically charged, insightful, and unwaveringly pragmatic. While he is no ordinary exemplar for a graduate student, I could not have been luckier nor happier to have had such a person as a Ph.D. adviser. He constantly challenged me to attain technical excellence and strive for a deeper clarity in my work. He fostered my curiosity and enabled me, as both a person and a scientist, to shine. I cannot thank him enough for all that he has done for me.

I have been so enriched by interactions with so many over the past five years. From the amazing folks at the Palomar and Keck observatories to the X-ray and γ-ray satellite builders to co-authors whom I have never met face-to-face, this thesis reflects the strengths and dedication of literally hundreds of people. Dr. Dale Frail, who I had the great fortune to work with on no less than 29 published refereed papers, has been one of the most important influences in my development. I will forever consider Dale a mentor, colleague, and a good friend.

Dale and Shri's original little 1996 afterglow duo blossomed into no less than a full-fledged team of a researchers. I am indebted to all of the past and existing members of the "Caltech–NRAO–CARA GRB Collaboration" who, without fail, graciously shared their talents and energies with me. The individuals who contributed to my thesis work are either co-authors on the respective published papers or explicitly mentioned in chapter-specific acknowledgments. Here, I particularly thank Prof. S. George Djorgovski, Re'em Sari, Prof. Jules Halpern (Columbia University), Alan Diercks, Chris Clemens, Steve Odewahn, Paul Price, Edo Berger, and Dan Reichart. Our collective adventures in the outer-limits of astronomy have been a source of immense pleasure and gratification.

My past and current officemates have been each extraordinarily influential. I thank Ken Banas, Ben Oppenheimer, Alice Shapley, Kurt Adelberger, and Rob Simcoe for their friendship, support, and knowledge. Over our frequent coffee and sushi breaks, I spent hundreds of hours talking and sharing ideas with Pieter van Dokkum and Dan Stern. I learned a great deal from Pieter and Dan about galaxies, cosmology, and data reductions; I hold them in the highest regard both personally and professionally. I shall never be able to fully articulate just how important my peers and colleagues have been.

For the past three years, I have been supported by a Fannie and John Hertz Foundation Graduate Fellowship. Each year, the Fellowship is supposed to be awarded to "America's most promising technical talents...who can be expected to have the greatest impact on the application of the physical sciences to human problems during the next half-century." I would never profess to be such a person but I certainly will strive attain that lofty goal. Indeed, it has been an honor to serve as Hertz Fellow.

I heartily thank my thesis committee members: Profs. Fiona Harrison (chair), Shri, George, Marc Kamionkowski, Nick Scoville. From the onset to the end, they consistently and thoughtfully encouraged me to build this thesis with an eye on the big picture. I thank the staff in the Astronomy department for their support and Josep Paredes at the University of Barcelona for his hospitality.

Even with such stimulating work environments, no one could be (or, at least, should be) truly fulfilled without close friendships. In this respect, I am very pleased to also count among my confidants and partners in pursuit of happy and full life: Michelle Brent, Josh Eisner, Gordon Squires, Ken B., Ben O., Geoff Criqui, John Ciorciari, Jared Bush, Jamie Miller, Jessica Rovello, Steve Agular, and Mark Galassi. These are the people who helped sustain me and who were always there through my highs and lows.

This thesis is dedicated to all my friends and those six people that are closest to my heart, without whom thesis would not have been possible nor worthwhile. Anna, Bec, Mom, Dad, Nanny, and Pop-pop, thank you.

Toward an Understanding of the
Progenitors of Gamma-Ray Bursts
by
Joshua Simon Bloom

Abstract

The various possibilities for the origin ("progenitors") of gamma-ray bursts (GRBs) manifest in differing observable properties. Through deep spectroscopic and high-resolution imaging observations of some GRB hosts, I demonstrate that well-localized long-duration GRBs are connected with otherwise normal star-forming galaxies at moderate redshifts of order unity. Using high-mass binary stellar population synthesis models, I quantify the expected spatial extent around galaxies of coalescing neutron stars, one of the leading contenders for GRB progenitors. I then test this scenario by examining the offset distribution of GRBs about their apparent hosts making extensive use of ground-based optical data from Keck and Palomar and space-based imaging from the *Hubble Space Telescope*. The offset distribution appears to be inconsistent with the coalescing neutron star binary hypothesis (and, similarly, black-hole–neutron star coalescences); instead, the distribution is statistically consistent with a population of progenitors that closely traces the ultra-violet light of galaxies. This is naturally explained by bursts which originate from the collapse of massive stars ("collapsars"). This claim is further supported by the unambiguous detections of intermediate-time (approximately three weeks after the bursts) emission "bumps" which appear substantially more red than the afterglows themselves. I claim that these bumps could originate from supernovae that occur at approximately the same time as the associated GRB; if true, GRB 980326 and GRB 011121 provide strong observational evidence connecting cosmological GRBs to high-redshift supernovae and implicate massive stars as the progenitors of at least some long-duration GRBs. Regardless of the true physical origin of these bumps, it appears that all viable alternative models of these bumps (such as dust scattering of the afterglow light) require a substantial amount of circumburst matter that is distributed as a wind-stratified medium; this too, implicates massive stars. Also suggested herein are some future observations which could further solidify or refute the supernova claim. In addition to the observational and modeling work, I also constructed the *Jacobs Camera* (JCAM), a dual-beam optical camera for the Palomar 200–inch Telescope designed to follow-up rapid GRB localizations.

Contents

Preface	1
1 Introduction and Summary	**3**
1.1 History, Phenomenolgy, and Afterglows	3
1.2 Proposed GRB Progenitor Scenarios	6
1.3 Summary of the Thesis: Constraining GRB Progenitors	9
1.3.1 Progenitor clues from the large-scale environments	11
1.3.2 An instrument to study the small-scale environments of GRBs	12
1.3.3 The stellar scale: connection to supernovae	14
I The Large-scale Environments of GRBs	**17**
2 The Spatial Distribution of Coalescing Neutron Star Binaries: Implications for Gamma-Ray Bursts	**19**
2.1 Introduction	19
2.2 Neutron Star Binary Population Synthesis	21
2.2.1 Initial conditions and binary evolution	21
2.2.2 Asymmetric supernovae kicks	21
2.3 Evolution of Binaries Systems in a Galactic Potential	22
2.4 Results	24
2.4.1 Orbital parameter distribution after the second supernova	24
2.4.2 Coalescence/birth rates	26
2.4.3 Spatial distribution	26
2.5 Discussion	27
2.6 Conclusions	29
3 The Host Galaxy of GRB 970508	**31**
3.1 Introduction	31
3.2 Observations and Analysis	32
3.3 Results	32
3.4 Discussion	33
3.5 Conclusions	35
4 The Host Galaxy of GRB 990123	**37**
4.1 Introduction	38
4.2 Observations and Data Reduction	38
4.3 Discussion	41
5 The Redshift and the Ordinary Host Galaxy of GRB 970228	**45**
5.1 Introduction	45

5.2	Observations and Reductions	47
5.3	The Redshift of GRB 970228	48
5.4	Implications of the Redshift	48
	5.4.1 Burst energetics	48
	5.4.2 The offset of the gamma-ray burst and the host morphology	50
	5.4.3 Physical parameters of the presumed host galaxy	51
	5.4.4 Star formation in the host	51
5.5	The Nature of the Host Galaxy	52
5.6	Discussion and Conclusion	52

6 The Observed Offset Distribution of Gamma-Ray Bursts from Their Host Galaxies: A Robust Clue to the Nature of the Progenitors   **57**

6.1	Introduction	58
6.2	Location of GRBs as a Clue to Their Origin	58
6.3	The Data: Selection and Reduction	59
	6.3.1 Dataset selection based on expected astrometric accuracy	61
	6.3.2 Imaging reductions	61
6.4	Astrometric Reductions & Dataset Levels	64
	6.4.1 Level 1: self-HST (differential)	64
	6.4.2 Level 2: HST→HST (differential)	64
	6.4.3 Level 3: GB→HST (differential)	65
	6.4.4 Level 4: GB→GB (differential)	66
	6.4.5 Level 5: RADIO→OPT (absolute)	66
6.5	Individual Offsets and Hosts	67
	6.5.1 GRB 970228	68
	6.5.2 GRB 970508	68
	6.5.3 GRB 970828	68
	6.5.4 GRB 971214	68
	6.5.5 GRB 980326	68
	6.5.6 GRB 980329	75
	6.5.7 GRB 980425	75
	6.5.8 GRB 980519	76
	6.5.9 GRB 980613	76
	6.5.10 GRB 980703	76
	6.5.11 GRB 981226	76
	6.5.12 GRB 990123	77
	6.5.13 GRB 990308	77
	6.5.14 GRB 990506	78
	6.5.15 GRB 990510	78
	6.5.16 GRB 990705	78
	6.5.17 GRB 990712	78
	6.5.18 GRB 991208	79
	6.5.19 GRB 991216	79
	6.5.20 GRB 000301C	79
	6.5.21 GRB 000418	79
6.6	The Observed Offset Distribution	80
	6.6.1 Angular offset	80
	6.6.2 Physical projection	84
	6.6.3 Host-normalized projected offset	84
	6.6.4 Accounting for the uncertainties in the offset measurements	85

6.7	Testing Progenitor Model Predictions	85
	6.7.1 Delayed merging remnants binaries (BH–NS and NS–NS)	85
	6.7.2 Massive stars (collapsars) and promptly bursting binaries (BH–He)	89
6.8	Discussion and Summary	91
6.A	Potential Sources of Astrometric Error	94
	6.A.1 Differential chromatic refraction	94
	6.A.2 Field distortion	94
6.B	Derivation of the Probability Histogram (PH)	95
6.C	Testing the Robustness of the KS Test	96

II The GRB/Supernova Connection 99

7 The Unusual Afterglow of the Gamma-ray Burst of 26 March 1998 as Evidence for a Supernova Connection 101

7.1	Introduction	102
7.2	The Unusual Optical Afterglow	102
7.3	A New Transient Source	107
7.4	The Supernova Interpretation	107
7.5	Implications of the Supernova Connection	108
7.A	Details of the Data Reduction for Table 7.1	109
	7.A.1 Photometric calibration	109
	7.A.2 Spectrophotometric measurement	110
	7.A.3 Photometry of the faint source	110
	7.A.4 Upper-limits	110
7.B	Notes on Figure 7.2	110
	7.B.1 Transient light curve	110
	7.B.2 Supernova light curve	110
7.C	Upper-limit Determination for Figure 7.1	111
7.D	Reduction Details for Figure 7.3	111
	7.D.1 Observing details	111
	7.D.2 Spectrophotometry	112

8 Detection of a supernova signature associated with GRB 011121 113

8.1	Introduction	114
8.2	Observations and Reductions	115
	8.2.1 Detection of GRB 011121 and the afterglow	115
	8.2.2 HST Observations and reductions	115
8.3	Results	115
8.4	Discussion and Conclusions	120

9 Expected Characteristics of the Subclass of Supernova Gamma-Ray Bursts 123

9.1	Introduction	123
9.2	How to Recognize S-GRBs	124
9.3	Application of Criteria to Proposed Associations	125
9.4	Discussion	127

III   An Instrument to Study the Small-scale Environments of GRBs — 131

10 JCAM: A Dual-Band Optical Imager for the Hale 200-inch Telescope at Palomar Observatory — 133
- 10.1 Introduction — 133
- 10.2 Scientific Motivation — 134
- 10.3 Instrumentation — 136
  - 10.3.1 Optical design — 136
  - 10.3.2 Mechanical design and construction — 143
  - 10.3.3 Electronics and software implementation — 143
- 10.4 Operations and Performance — 144
  - 10.4.1 Initiating operations — 145
  - 10.4.2 Data taking and observing procedures — 145
  - 10.4.3 Preliminary results — 147
  - 10.4.4 Deficiencies — 150
  - 10.4.5 Future extensions — 151

11 Epilogue and Future Steps — 155
- 11.1 On the Offset Distribution of GRBs — 155
- 11.2 Re-examining the GRB–Supernova Connection — 157
  - 11.2.1 S-GRBs — 157
  - 11.2.2 Supernova bumps — 159
- 11.3 Conclusions: What Makes Gamma-ray Bursts? — 162

Bibliography — 164

List of Figures

1.1	Anatomy of a gamma-ray burst explosion	5
1.2	The redshift and energy distributions of the 22 cosmological GRBs with known redshift	7
1.3	Schematic of plausible theoretical scenarios for the progenitors of classic GRBs	10
2.1	The distribution of orbital parameters after the second supernovae for bound NS–NS pairs	23
2.2	The distribution of merger times after second supernovae as a function of system velocity	25
2.3	The radial distribution of coalescing neutron stars around galaxies of various potentials	27
2.4	NS–NS merger rate dependence on redshift	28
3.1	Light curve of the optical transient of GRB 970508	33
3.2	The weighted average spectrum of the host galaxy of GRB 970508	34
4.1	Three epochs of Keck I K-band imaging of the field of GRB 990123	39
4.2	The HST/STIS drizzle image of the field of GRB 990123	40
5.1	The weighted average spectrum of the host galaxy of GRB 970228, obtained at the Keck II telescope	49
5.2	A $3'' \times 3''$ region of the HST/STIS image of the host galaxy of GRB 970228	50
5.3	Comparison of the color-magnitude of the host galaxy of GRB 970228 with the Hubble Deep Field-North	53
5.4	Median-binned portion of the host spectrum near the Balmer decrement	54
6.1	Example Keck and HST/STIS images of the field of GRB 981226	65
6.2	The location of individual GRBs about their host galaxies	70
6.2	The location of individual GRBs about their host galaxies (cont.)	71
6.2	The location of individual GRBs about their host galaxies (cont.)	72
6.2	The location of individual GRBs about their host galaxies (cont.)	73
6.2	The location of individual GRBs about their host galaxies (cont.)	74
6.3	The angular distribution of 20 gamma-ray bursts about their presumed host galaxy	80
6.4	Host-normalized offset distribution	81
6.5	The GRB offset distribution as a function of normalized galactocentric radius	82
6.6	The cumulative GRB offset distribution as a function of host half-light radius	88
6.7	Offset distribution of GRBs compared with delayed merging remnant binaries (NS–NS and BH–NS) prediction	90
6.8	GRB Offset distribution compared with host galaxy star formation model	93
6.9	Geometry for the offset distribution probability calculation in Appendix 6.B	97
6.10	Example offset distribution functions $p(r)$	98
7.1	The R-band light curve of the afterglow of GRB 980326	104
7.2	Images of the field of GRB 980326 at three epochs	105
7.3	The spectra of the transient on March 29.27 and April 23.83 1998 UT	106

8.1 *Hubble Space Telescope* image of the field of GRB 011121 116
8.2 Light curves of the intermediate-time red bump of GRB 011121 118
8.3 The spectral flux distributions of the red bump at the time of the four HST epochs ... 119

9.1 The γ-ray light curves of GRB 980425 and other potential S-GRBs 126

10.1 Theoretical evolution of the reverse-forward shock transition in the early afterglow ... 135
10.2 Picture of JCAM mounted at the East Arm 137
10.3 The calculated filter response curves of JCAM 140
10.4 Schematic of the hardware and software configuration of JCAM 144
10.5 Screenshot of the JCAM graphical user interface 146
10.6 JCAM images of the afterglow of GRB 010222 148

11.1 Update to chapter 6 of new offsets and *HST* images of host galaxies. 156
11.2 A future step toward resolving the progenitor question: space-based spectroscopy of intermediate-time emission components 161

List of Tables

1.1	Estimated Rates of GRBs and Plausible Progenitors	8
2.1	The Spatial Distribution of Coalescing Neutron Star Binaries in Various Galactic Potentials	24
2.2	The Bound NS–NS Binary Birthrate and Merger Time Properties as a Function of Supernova Kick Strength	26
3.1	Late-time GRB 970508 Imaging and Spectroscopic Observations	36
6.1	GRB Host and Astrometry Observing Log	60
6.1	GRB Host and Astrometry Observing Log	62
6.2	Measured Angular Offsets and Physical Projections	63
6.3	Host Detection Probabilities and Host Normalized Offsets	69
6.4	Comparison of Observed Offset Distributions to Various Progenitor model Predictions	87
7.1	Keck II Optical Observations of GRB 980326	103
8.1	Log of HST Imaging and Photometry of the OT of GRB 011121	117
9.1	GRB/Supernovae Associations: Which are Truly S-GRBs?	127
10.1	Properties of Optical Elements in JCAM	138
10.2	Summary of JCAM Filter Properties	141
10.3	Synthetic Magnitudes of Primary Standard Stars Through the JCAM Filter Set	142
10.4	Summary of GRB Triggers Observed to Date with JCAM	143
10.5	JCAM Detector Characteristics	148
10.6	JCAM Filter Response Curves	152
10.6	JCAM Filter Response Curves	153
10.6	JCAM Filter Response Curves	154
11.1	Summary of Proposed Cosmological GRBs with Associated Supernovae	158
11.2	Summary Assessment of Progenitor Scenarios	163

Preface

But the reason so many of you live, work and study here is that there are so many more questions yet to be answered...And so I wonder,...Are we alone in the universe? **What causes gamma ray bursts?** What makes up the missing mass of the universe? What's in those black holes, anyway? And maybe the biggest question of all: How in the wide world can you add $3 billion in market capitalization simply by adding .com to the end of a name?

William Jefferson Clinton
42nd President of the United States of America
Science and Technology Policy Speech
21 January 2000, Caltech

These ponderances, spoken just before the bursting of the Internet "bubble," invigorated me. Rarely, I suspect, does a sitting president publicly ask the question that is the central endeavor of one's PhD thesis. But, given the topic and the timing—gamma-ray bursts, during one of the most enlightening periods of understanding of the phenomenon—perhaps we should not be surprised that this happened. Of course, the cynical view is that the President's speech writers had scoured the Caltech web site before his visit and encountered a number of the astronomy press releases that had been generated here over the years, drawing up a number of relevant questions for a Caltech-specific audience. The idealist view, one that is perhaps more comforting to accept, is that the question of what makes gamma-ray bursts really is on the minds not just of a few astronomers but on a much larger audience.

For those that are familiar with the phenomenon, this public appeal of gamma-ray bursts is almost assured—GRB descriptions are, after all, awash in superlatives. We now know, for instance, that GRBs are one of the *brightest* events in the universe, briefly reaching luminosities comparable to the integrated luminosity of a few hundred thousand galaxies. The bursts are some of the *rarest* well-studied transient events (only a few per galaxy per 10 million years) and probably represent the violent death and/or birth of the *most dense* objects known, namely black holes and neutron stars. Because of the extreme densities and accelerations of the masses involved in triggering a GRB, the events leading up to a GRB explosion holds the greatest promise of impulsively releasing as yet undetected gravitational waves.

That this central question—What causes gamma-ray bursts?—can even be asked now with a straight face is nothing short of remarkable. Not too long ago—pre-1997, to be precise—no one knew for sure from where GRBs originated. Was the origin of the bursts that of primordial anti-matter comets smashing into the Oort cloud (distance of scale of 100 pc; Dermer 1996)[1] or the re-connection of superconducting cosmic strings (distance of scale of 10^{10} pc; Paczyński 1988) or some place more conventional (such as in our Galaxy or more distant galaxies)?

This remarkable ignorance, let alone the lack of any solid connection between GRBs and other known astrophysical entities, persisted for 29 years and 10 months after the first detection of a GRB. Then, in May 1997, shortly following the detection of the first long-lived emission following a GRB ("afterglow"), finally the first step was taken. An optical spectrum of the afterglow obtained at the Keck telescopes revealed the burst to have originated from at least a redshift of $z = 0.835$, proving that at least one GRB originated from "cosmological" distances.

This was the state of affairs when I began my PhD thesis at Caltech in the fall of 1997—just one GRB was known to be of a cosmological origin and a total of two afterglows had been discovered. My early interest in afterglow follow-up work stemmed from the belief that the intense study of afterglows would surely lead to physical insights about the nature of the afterglows themselves and GRB progenitors. Indeed, with over 30 GRB afterglows discovered and studied in the past five years, the community

---

[1] 1 parsec = 3.085×10^{18} cm or 3.27 light-years

has solidified GRBs as of a cosmological origin and learned a great deal about the physical processes underlying afterglows. My thesis, however, focuses on the later issue, uncovering the progenitors.

The ultimate conclusion of this work—that most GRBs probably arise from the death of massive stars rather than the coalescences of massive compact binary stars—was rather unexpected since the predominant view in 1997 was that the latter objects were likely responsible for such bursts. The connection between GRBs and massive stars represents, as van Paradijs et al. (2000) have also pointed out, a harmonious and beautiful full-circle revelation: the first theory of gamma-ray bursts, posited by Colgate (1968) before GRBs had even be discovered, suggested that GRBs could arise during a supernova explosion.

Notes on the contents of this thesis: Thanks to the relative ignorance of the physics of GRBs and the small wavelength regime in which GRBs and their aftermath had been observed (X-rays to GeV gamma-rays), the introduction to a PhD thesis ten or even five years ago could realistically have captured the sum-total knowledge of the phenomena for the reader. Yet, after 5 heady years in the afterglow era, the information explosion[2] would surely require hundreds of pages for a comprehensive exposition of GRBs (think of writing a comprehensive review on galaxy evolution or supernovae).

This is not my intention with the scope of the following summary chapter; instead, I give a brief overview of the observations and current theoretical understanding of the phenomena[3]. Then I present some of the more salient topics related to progenitors and their host galaxies, drawing specific attention to the work in body of the thesis. For a more thorough review of the current state of the field, the reader is referred to Fishman & Meegan (1995) (GRB science pre-1997), Piran (1999) (fireball and afterglow physics), van Paradijs et al. (2000) (afterglow observations), and Fryer et al. (1999a) (progenitor models).

The next eight chapters were prepared for a total of five different journals, each, as such, with differing presentation and referencing styles. The audience for each article varied as well—the *Nature* article (chapter 7), for instance, was geared to a general science audience, while the *PASP* article (chapter 10) was written primarily for an astronomical instrumentation audience. In the interest of continuity, I have homogenized the referencing style and nomenclature from chapter to chapter.

In some chapters, I have also included some additional text and figures that were necessarily cut from the published article due to space constraints. In chapter 9, for instance, I include example γ-ray light curves of possible supernova–GRBs, and in chapter 7 I provide an expanded explanation of some of the data reduction methods. Looking back, some ideas in the chapters proved to be less salient (and correct!) than others and so there was the temptation to jettison the chaff. Since this thesis is comprised mostly of published articles that themselves (should) reflect the progression of the field, I have, however, tried to steer clear of constructing such a revisionist history. As such, all of the "new" ideas augmented to the published versions appeared first in submitted versions of the paper. Chapter 11 is designated as forum for redresses and epilogues to the body of this thesis.

---

[2] As of mid-2001, GRBs entered the literature at a rate of 1.5 per day, 50% higher than the rate in 1994 (Hurley 2002).

[3] The first section of chapter 1 contains some of the text in the published version of chapter 6. The introduction of chapter 6 is commensurately abridged.

CHAPTER 1

Introduction and Summary

SECTION 1.1
History, Phenomenolgy, and Afterglows

Gamma-ray bursts, otherwise extinguished by the Earth's atmosphere, were discovered serendipitously (Klebesadel et al. 1973; Strong et al. 1974) by space-based US satellites designed to insure compliance with the Limited Nuclear Test Ban Treaty (signed July 25, 1963) by searching for the γ-ray emission that accompanies nuclear weapons testing. In gross properties, the 23 bursts presented in the discovery papers were not unlike the some 3000 observed to date by the many GRB-specific satellites that followed. One notable sub-class of the gamma-ray burst (GRB) phenomenon are the so-called "Soft Gamma-Ray Repeaters" (SGRs) which, since the late-1970s, have been definitively associated with highly-magnetized isolated neutron stars (magnetars) in our Galaxy or the Large Magellanic Cloud (LMC) (see Harding 2001 for review).

The duration of "classic" GRBs (i.e., those GRBs which are not SGRs), observationally determined as the time that the flux exceeds some threshold above the sky background level, ranges from a few milliseconds to thousands of seconds (e.g., Kouveliotou et al. 1996). The peak of the spectral energy distribution of bursts falls in the range of ~ 50–1000 keV (Mallozzi et al. 1995) but both ends of this range likely exist due to the trigger inefficiencies of GRB satellites (see, e.g., fig. 3 of Lloyd & Petrosian 1999; although see Brainerd 1998).[1]

GRBs seem to occur at random times and from random locations in the sky, about 4 times per day at current detector thresholds. Though it was known that the sky distribution of GRBs appeared roughly isotropic for years, the *Burst and Transient Source Experiment* (BATSE), the prime workhorse of GRB astronomy in the early- to mid-1990s, placed the strongest constraints showing GRBs to be isotropically distributed to within a high degree of confidence (Meegan et al. 1992; see also table 8 of Paciesas et al. 1999).

Though, by 1995, no classical GRB had been definitively connected with any other astrophysical entity, the observed isotropy was taken by many as evidence for a cosmological progenitor origin. The paucity of faint bursts relative to the number expected if the bursts originated homogeneously in Euclidean space also served as evidence for a cosmological origin (e.g., Fenimore et al. 1993; Fenimore & Bloom 1995). During the "Great Debate" on the distance scale to GRBs (Nemiroff 1995), Paczyński (1995) argued for the cosmological origin of GRBs based primarily on these two points while Lamb (1995) explained how the same data were consistent with a galactic progenitor origin.[2] I attended the

---

[1] Until very recently, a burst was not considered a GRB until its energy spectrum peaked above ~ 30 keV. It is the adherence to this definition of GRBs that may be restricting a deeper insight into the nature of GRBs. If a burst is found to peak at lower-energies (a so-called X-ray Flash; XRF), but retains most of the other properties of classic GRBs, then it may simply be a GRB at a high redshift (one of remaining holy-grails of observational GRB astronomy). Heise et al. (2001) has recently written a nice admonition to the community about classification of such phenomena.

[2] That two such contrasting views on the same data existed should remind us of the quote from Antonio in the

Debate and thought that D. Lamb made a persuasive argument for Galactic scenarios despite the small theoretical parameter space then allowed by the isotropy and brightness distribution observations.

The main impedance to progress was the difficulty of localizing bursts to an accuracy high enough to unequivocally associate an individual GRB with some other astrophysical entity. Several concerted efforts were made to find counterparts (e.g., Schaefer et al. 1987; Vanderspek et al. 1994; Vrba et al. 1995; Frail & Kulkarni 1995; Greiner 1995), but we now know that such surveys were either too shallow or too delayed in time to catch rapidly fading GRB counterparts. Aside from the hope that GRBs would produce/induce emission at some other frequency, some counterpart searches were strongly motivated by (prescient) theoretical predictions for the existence of lower-frequency counterparts (Paczyński & Rhoads 1993; Mészáros & Rees 1993; Katz 1994; Mészáros et al. 1994). These "afterglow" models were a natural consequence of the (also prescient) predictions for cosmological GRBs from highly-relativistic outflow of a low-baryon-loaded fireball (Paczyński 1986; Goodman 1986; Mészáros & Rees 1992; Mészáros & Rees 1993).

In large measure the localization problem was due to both the transient nature of the phenomena and the fact that the incident direction of γ-rays are difficult to pinpoint with a single detector; for example, the typical 1-σ uncertainty in the location of a GRB using the *Burst and Transient Source Experiment* (BATSE) was 4–8 degree in radius (Briggs et al. 1999). The Interplanetary Network (IPN; see Cline et al. 1999) localized GRBs using burst arrival times at several spacecrafts throughout the solar system and provided accurate localizations (3 σ localizations of ∼few to hundreds × sq. arcmin) to ground-based observers; however, the localizations were reported with large time delays (days to months after the GRB).

The crucial breakthrough came in early 1997, shortly following the launch of the *BeppoSAX* satellite (Boella et al. 1997). On-board instruments (Frontera et al. 1997; Jager et al. 1997) were used to rapidly localize the prompt and long-lived hard X-ray emission of the GRB of 28 February 1997 (GRB 970228) to a 3 σ accuracy of 3 arcmin (radius) and relay the location to ground-based observers in a matter of hours. Fading X-ray (Costa et al. 1997a) and optical (van Paradijs et al. 1997) emission (afterglow) associated with GRB 970228 were discovered. Ground-based observers noted (Metzger et al. 1997c; van Paradijs et al. 1997) a faint nebulosity in the vicinity of the optical transient (OT) afterglow. Subsequent *Hubble Space Telescope* (HST) imaging resolved the nebulosity (Sahu et al. 1997) and showed that the morphology was indicative of a distant galaxy (Sahu et al. 1997). We now know the redshift of this faint, blue galaxy is $z = 0.695$ (chapter 5, Bloom et al. 2001a).

The next prompt localization of a GRB yielded the first measured distance to a GRB through optical absorption spectroscopy: GRB 970508 occurred from a redshift $z \geq 0.835$ (Metzger et al. 1997b). The first radio afterglow was detected from GRB 970508 which, through observations of scintillation, led to the robust inference of super-luminal motion of the GRB ejecta (Frail et al. 1997; see below). These measurements (along with the dozen other redshifts now associated with individual GRBs) have effectively ended the distance scale debate and solidified GRBs as one of the most energetic phenomena known (see Kulkarni et al. 2000; Frail et al. 2001). One of the most remarkable aspects of GRB science (e.g., Wijers et al. 1997) was the enormous success, as evidenced by the observations of GRB 970228 and GRB 970508, of the theoretical predictions for the existence and behavior of GRB afterglows (Paczyński & Rhoads 1993; Mészáros & Rees 1997a; Vietri 1997).

The cosmological nature of GRBs now frames our basic understanding of the physics of GRB phenomena. The general energetics are well-constrained: given the observed fluences and redshifts, approximately 10^{51}–10^{53} erg in γ-ray radiation is released in a matter of a few seconds in every GRB (see fig. 1.2). The GRB variability timescale suggests that this energy is quickly deposited by a "central engine" in a small volume of space (radius $r \lesssim 1000$ km) and is essentially optically thick to γ-ray radiation at early times. This opaque fireball of energy (see below) then expands adiabatically and relativistically until the γ-ray radiation can escape; there, the GRB is thought to arise from the interaction of internal shocks initiated by the central engine (e.g., Fenimore et al. 1999).

Merchant of Venice, Act I,iii: "Mark you this, Bassanio, The devil can cite Scripture for his purpose."

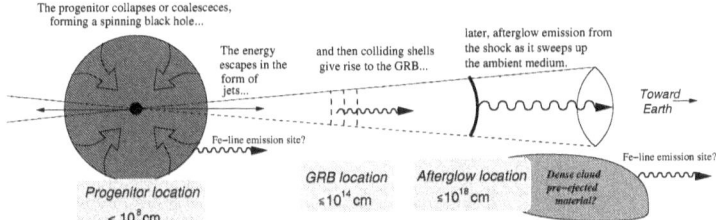

Figure 1.1 Anatomy of a gamma-ray burst explosion. The dark circle represents the newly formed spinning black hole at the center of an imploding star (or merging compact binary system). The long-lived afterglow emission that we see arises from the swept-up material; in this material, relativistic electrons radiate sychrotron light in an amplified magnetic field. Due to the extreme velocity of the jet, the whole sequence of events is compressed in time as viewed from Earth.

The short variability timescale of a GRB and high total energy release would tend to imply that the optical depth to pair-production at the explosion site is exceedingly high ($\tau_{\gamma\gamma} \gtrsim 10^{12}$) yet GRB spectra are optically thin. It was recognized in the mid-1980s that this so called "compactness problem" can be avoided by invoking relativistic motion (Goodman 1986; Paczyński 1986). If the surface of emission of γ-rays is moving toward the observer at a bulk Lorentz factor Γ, then the optical depth to pair-production is reduced due to two effects. First, the fraction of photons that can pair-produce in the frame of the moving surface is lower than inferred by an outside observer (by Γ^2 for a flat spectrum), who measures a blue-shifted spectrum. Second, special relativistic effects allow for the emission radius to be larger by Γ^2 than the variability estimate (about 10^{14} cm rather than 10^8 cm). By requiring that $\tau_{\gamma\gamma} < 1$, the source of GRB emission must be moving with $\Gamma \gtrsim 100$ at the time of the GRB.

The coupling of the fireball energy to any entrained baryons will tend to stall the outward expansion of the flow. Specifically, if the total energy in the fireball is E_0 ($\approx 10^{51}$ erg), then the total amount of baryonic mass allowed in the flow is $M = E_0/\Gamma c^2 \lesssim 10^{-5} M_\odot$. The elegant solution to the compactness problem, then, places an important constraint on the nature of GRB ejecta (see also Piran 1999): the fireball must be nearly devoid of baryons.

After the GRB, the relativistic blastwave continues its outward expansion and begins to sweep up the ambient medium. Taking $n = 1$ cm^{-3} to be the density of the surrounding medium, the blastwave begins to slow considerably by a radius $R \approx (E_0/4 m_H c^2 \Gamma^2)^{1/3} \approx 10^{16}$–$10^{17}$ cm and some of the kinetic energy is then converted into internal motion within the shock (Mészáros & Rees 1993). Here m_H is the mass of a hydrogen atom. The transient afterglow phenomenon, thought to arise at this radius, is likely due to synchrotron radiation arising from the interaction of the relativistic ejecta and the ambient medium surrounding the burst site (see van Paradijs et al. 2000; Kulkarni et al. 2000; Djorgovski et al. 2001, for reviews).

Indirectly, the initial Γ of some GRBs have been constrained in the context of the compactness problem and early-time observations of afterglows (e.g., Mészáros et al. 1993; Sari & Piran 1999a; Ramirez-Ruiz & Fenimore 2000; Soderberg & Ramirez-Ruiz 2002). By noting an abrupt quenching of scintillation behavior a few days after GRB 970508, Frail et al. (1997) showed that the afterglow emitting region must have grown with apparent superluminal speeds, observationally solidifying relativistic motion as a fundamental property of GRBs.

There have now been tentative detections of transient X-ray line features in five GRB afterglows (e.g., 970508 and 970828 Piro et al. 1999; Yoshida et al. 1999). The most convincing detection so far comes from observations of the afterglow of GRB 991216 (Piro et al. 2000). Individually, the observational significance of the line detections are marginal, but on the whole there appears to be a good case for line emission features in the afterglow of some GRBs. If so, there must exist dense matter in the vicinity of the explosion (e.g., Weth et al. 2000; Vietri et al. 1999; Lazzati et al. 2000). Mészáros & Rees (2001) have also suggested that the Fe lines may be produced if some of the waste energy from the central engine heats up a "bubble" of matter from the progenitor, a natural consequence of jets propagating in a dense stellar interior. The bubble breaks out from the progenitor remnant on a timescale of hours to days. Figure 1.1 depicts the relative locations of the suggested sources of the X-ray line emission from a generalized progenitor.

Like most other high-energy phenomena, it is now widely accepted that gamma-ray burst emission is collimated (or "jetted"). Observationally, jetting should be manifested as a (variable) polarization signal in the afterglow (Ghisellini & Lazzati 1999; Sari 1999); such signatures have been detected in some GRB afterglows (Covino et al. 1999; Wijers et al. 1999; Björnsson & Lindfors 2000; Rol et al. 2000). The observed evolution of GRB afterglows, often showing a break in the light curves from 0.5–50 days, also appear to conform to basic predictions from the dynamics of jetted outflows (Rhoads 1999; Sari et al. 1999). Aside from relaxing the overall energy requirements, the establishment of jetting in GRBs also implies that the true rate of GRBs in the universe is substantially higher than previously believed (by a factor of $\sim$550, Frail et al. 2001).

SECTION 1.2
Proposed GRB Progenitor Scenarios

While the GRB emission and the afterglow phenomenon of long-duration bursts are now reasonably well-understood, one large outstanding question remains: what makes a gamma-ray burst? Specifically, what are the astrophysical objects, the "progenitors," which produce GRBs? Those theoretical progenitor scenarios which have remained feasible in the afterglow era are principally constrained by the following considerations:

- The implied (isotropic) energy release in γ-rays are typically 10^{-3}–10^{-1} times the rest-mass energy of the Sun. The estimated efficiency of conversion of the initial input energy (either Poynting flux or baryonic matter) to γ-rays ranges from $\sim$1% (e.g., Kumar 1999) to as much as $\sim$60% (e.g., Kobayashi & Sari 2001); therefore, the best-guess estimate of the total energy release (including neutrino and gravitational-wave losses) is roughly comparable to the rest-mass energy of one solar mass.

- The GRB variability timescale (few ms) observed implies that the energy deposition takes place in a small region of space (radius of $c \times 1$ ms ≈ 300 km). The range in total burst durations suggest that the central engine must live for less than one second and up to thousands of seconds.

- The inferred rate of GRB occurrence (table 1.1) and the lack of burst repetition (e.g., Hakkila et al. 1998) suggest that GRB events are rare and catastrophically destroy the individual progenitors.

The progenitor scenarios which most naturally explain these observables fall in to three broad classes—the coalescence of binary compact stellar remnants, the explosion of a massive star ("collapsar"), and the accretion-induced collapse of a differentially rotating compact object ("DRACO"; Kluźniak & Ruderman 1998). An active galactic nucleus (AGN) origin is another possibility, whereby a main-sequence (MS) or white-dwarf (WD) is tidally disrupted near a super-massive black hole. In such scenarios, however, the variability timescale still requires the energy source to be stellar-mass objects (Carter 1992; Cheng & Wang 1999).

Here, I briefly summarize the popular progenitor models and refer the reader to Fryer et al. (1999a) for a more in-depth review of the black-hole accretion disk progenitors models. Figure 1.3 depicts a schematic compilation of the major progenitor scenarios for classic GRBs along with references for the

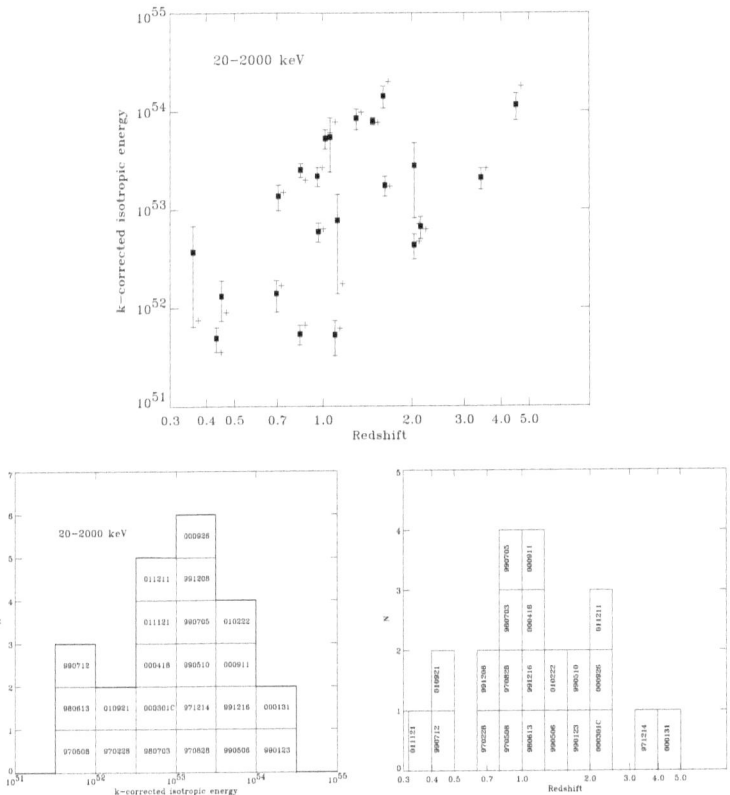

Figure 1.2 The redshift and energy distributions of the 22 cosmological GRBs with known redshift. At top, the k-corrected prompt energy release versus redshift in the restframe 20–2000 keV bandpass assuming isotropic emission. The 2 σ error bars are shown as well as the derived energies if no cosmological k-correction is applied (denoted as a cross "+"). At bottom left, the histogram of k-corrected GRBs energies. At bottom right, the observed redshift distribution of GRBs with measured redshifts. Note that while this observed distribution is reflective of the true GRB rate as a function of redshift, this is not, in general, the true GRB rate: for example, owing to a lack of strong star formation lines at observer-frame optical wavebands, there is a strong selection against finding emission-line redshifts in the redshift range $1.7 \lesssim z \lesssim 2.5$. Figures adapted and updated from Bloom et al. (2001b).

Table 1.1. Estimated Rates of GRBs and Plausible Progenitors

Progenitor/ Phenomenon	Rate (yr^{-1} Gpc^{-3}) local rate ($z = 0$)	Ref.
NS–NS	80	Phinney (1991)
BH–NS	10–300	Fryer et al. (1999a)
BH–WD	10	Fryer et al. (1999a)
BH–He	1000	Fryer et al. (1999a)
Type Ib/Ic	6×10^4	Phinney (1991)
GRBs	0.5^a	Schmidt (2001); Wijers et al. (1998)
	250^b	Frail et al. (2001)

^{a}Rates not including beaming. Assumes that GRBs follow the star-formation rate in the universe.
^{b}Rates including the effects of beaming.

Note. — Aside from the rate of Type Ib/Ic SNe events, the rates of GRBs and possible GRB progenitors are uncertain by at least a factor of two. In the case of NS–NS and BH–NS mergers, the true rates probably are uncertain by at least an order of magnitude (e.g., Kalogera et al. 2001a). All of the progenitor scenarios listed closely scale with the rate of star formation; therefore, the rates at redshift of $z = 1$ are a factor of ∼ten higher than locally.

various theoretical treatments of each scenario. The time sequence for the (supposed) predominant production channel for each family of progenitor scenarios is shown, although there are variants for each family that could plausibly produce the same trigger. For instance, only binary progenitors for merging scenarios are depicted, but some mergers may occur after stellar capture in dense cluster cores (e.g., Sigurdsson & Rees 1997).

A spinning BH is formed in both the collapsar and the merging remnant class of progenitors. The debris, either from the stellar core of the collapsar or a tidally disrupted neutron star, forms a temporary accretion disk (or "torus") which then falls into the BH releasing a fraction of gravitational potential energy of the matter. In this general picture (see Rees 1999, for a review), the lifetime of the accretion disk accounts for the duration of the GRB and the light-crossing time of the BH accounts for the variability timescale. The GRB is powered by the energy extracted either from the spin energy of the hole or from the gravitational energy of the in-falling matter.

The coalescing compact binary class (Paczyński 1986; Goodman 1986; Eichler et al. 1989) was favored before the first redshift determination because the existence of coalescence events of a double neutron star binaries (NS–NS) was observationally assured: at least a few known NS–NS systems in our Galaxy (e.g., PSR 1913+16, PSR 1534+12) will merge in a Hubble time thanks to the gravitational radiation of the binary orbital angular momentum (see Taylor 1994). Further, the best estimate of the rate of NS–NS coalescence in the Universe (e.g., Phinney 1991; Narayan et al. 1992) was comparable to an estimate of the GRB rate (e.g., Fenimore et al. 1993)[3]. Recently, stellar evolution models have suggested that black hole–neutron star binaries (BH–NS) may be formed at rates comparable to or even higher than NS–NS binaries (e.g., Bethe & Brown 1998), though no such systems have yet been observed. There are other merging remnant binaries which may form GRBs, notably merging black

---
[3] The latter estimate assumed a constant bursting rate as a function of redshift and that the faintest bursts only had been detected to redshift of unity. These two assumptions, which proved to be incorrect, upwardly biased the local unbeamed estimate by ∼150 (Wijers et al. 1998).

hole–white dwarf (BH–WD) binaries (Tutukov & Yungelson 1994; Fryer et al. 1999b) and black hole–helium star binaries (BH–He) (Fryer & Woosley 1998). Table 1.1 provides a summary of the various rates estimates of some of these GRB progenitors.

The collapsar class is comprised of a rotating massive Wolf-Rayet star, either isolated or in a binary system, whose iron core subsequently collapses directly to form a black hole. The basic picture of a collapsar (or "failed Type Ib supernova," as it was often called) was pioneered by Woosley (1993). To avoid baryon loading the progenitor star should have lost most, if not all, of its extended gas envelope of hydrogen by the time of collapse. The progenitors of collapsars—likely Wolf-Rayet stars—are then closely related to the progenitors of hydrogen-deficient supernova, namely type Ib/Ic supernovae (see MacFadyen & Woosley 1999).

As can be seen in table 1.1, clearly not all type Ib/Ic supernovae can be accompanied by a GRB, even if beaming is taken into account. Perhaps one distinguishing difference is that high angular momentum is necessary in collapsars. High angular momentum centrifugally supports a transient torus around the BH, fostering an extended timescale (∼tens of seconds) for mass-energy injection. Further, angular momentum creates a natural rotation axis that sets up large density gradients which then allow for the expanding blastwave to reach relativistic speeds. The efficiency of energy conversion is also helped around a spinning BH for two reasons. First, the innermost stable orbit around a Kerr BH is smaller than for a non-rotating BH, allowing for more gravitational potential of the accretion torus to be tapped. Second, rotational energy extraction from the hole becomes possible via the Blandford-Znajek process. Given the benefits of high angular momentum in the collapsar scenario, it is thus reasonable to suggest that collapsars might be more readily formed in close binary systems.

The accretion-induced collapse scenarios posit that the energy for a gamma-ray burst is stored in the rotation and/or magnetic field of a compact object. Usov (1992) first suggested that cosmological GRBs could be powered by dipole radiation from a magnetized accreting white dwarf which collapses to form a neutron star. Kluźniak & Ruderman (1998) suggested a variant to this by noting that differential rotation is temporarily induced when an object collapses due to accretion (either a white dwarf collapsing to a neutron star or a neutron star collapsing to a black hole). During the differential-rotation stage, a buoyant magnetic field dissipates the rotational energy and could create a brief episodic burst of electromagnetic energy. Though the timescales for the energy dissipation by such a mechanism appear to be plausible for GRBs, for many years DRACO and magnetar models were not favored since the energy reservoir was considerably less than one solar mass. However, after the relatively recent firm establishment of jetting in GRBs, thus reducing the overall energy requirements, such models might now be more viable than previously believed.

SECTION 1.3
Summary of the Thesis: Constraining GRB Progenitors

Like supernovae, the progenitors of gamma-ray bursts have become a subject of intense interest and study for many decades. But unlike observed supernovae, GRBs are more distant and occur much less often. This implies that the systematic study of GRB progenitors must progress without the aid of nearby examples, where by assumption the study of such examples could be done with high photometric and astrometric precision. Indeed, unlike as has already been seen with a few supernovae (e.g., SN 1987a as a blue-supergiant; White & Malin 1987; Walborn et al. 1989), there is little hope that we will ever have a resolved pre-discovery image of a GRB progenitor.

We are thus forced to uncover the progenitors of GRBs by indirect means, and it is the thrust of my thesis work to do so. This thesis is separated into three parts corresponding to different aspects of attack on the GRB progenitor question, progressing from the very large scale (galaxies and locations within galaxies) to the very small scale (the stellar scale). In the first part, entitled "The Large-scale Environments of GRBs," I posit that observations the host galaxies of GRBs place important constraints on the nature of GRB progenitors. I show in chapter 2 that it is possible to distinguish between various

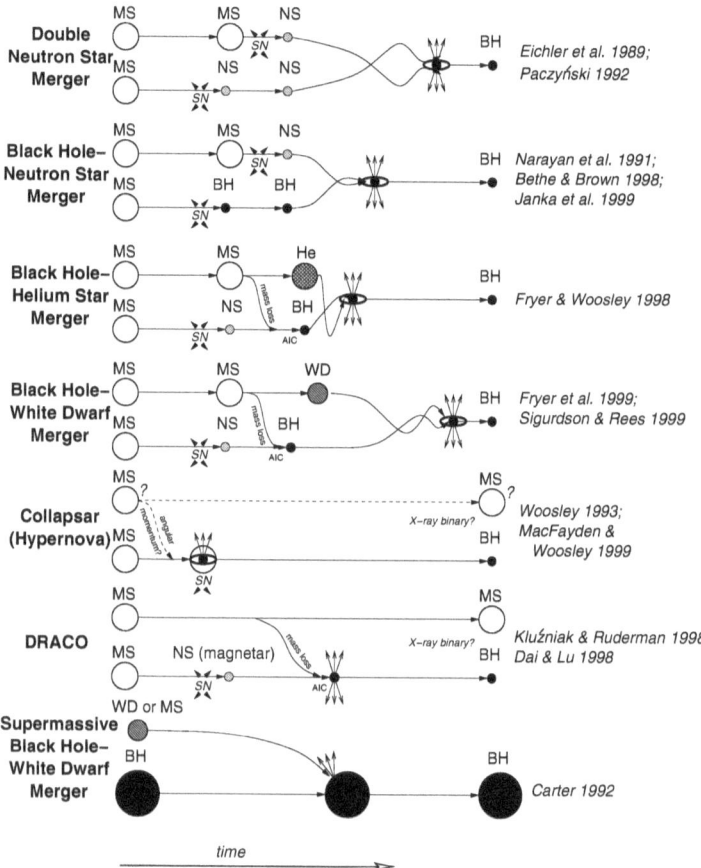

Figure 1.3 Schematic of plausible theoretical scenarios for the progenitors of classic gamma-ray bursts. In merger scenarios, the primary star (more massive at ZAMS) is depicted as the bottom component. The dominant production channel for each scenario is shown. The (rough) relative in-spiral time due to gravitational radiation for the four scenarios at top are shown (e.g., BH–He mergers occur, in general, much more rapidly than NS–NS or NS–WD mergers). AIC = "accretion-induced collapse"; SN = supernova explosion.

progenitor models by observing the distribution of GRBs around their host galaxies. In chapter 6, I present a comprehensive observational study of the distribution of GRBs around galaxies which, in the context of chapter 2, provides one of the strongest constraints to-date on the nature of the progenitors.

In part two of this thesis, entitled "The GRB/Supernova Connection," I present the first observational evidence for bright intermediate-time emission in a GRB afterglow. I interpret this light curve "bump" as due to a supernova which occurred contemporaneously with a distant GRB. Then, in light of the observations of a very nearby supernova with a probable connection to a GRB, I propose a new sub-classification of GRBs based on the physical underpinnings of a true GRB-supernova connection.

In part three of this thesis, entitled "An instrument to study the small-scale environments of GRBs," I describe the design and construction of the *Jacobs Camera* (JCAM) for the Palomar 200 inch Telescope.

1.3.1 Progenitor clues from the large-scale environments

The largest scale: Studies of GRBs on the gigaparsec scale offer some big clues to the nature of the progenitors. First and foremost, of course, is the detection of redshifted absorption lines which immediately renders all Galactic models for GRBs untenable. The relatively moderate redshift distribution of GRBs (fig. 1.2) also suggests that some of the more exotic progenitor scenarios at high redshifts, such as bursts from super-conducting cosmic strings (Paczyński 1988), are incorrect. The remainder of the viable models all posit a progenitor birthsite (though not necessarily explosion site) near to and as part of the stellar mass of galaxies.

The connection between GRBs and stars is borne out by the observed redshift distribution (fig. 1.2). Qualitatively, the rate of GRBs appears to peak around redshift of unity, just as the inferred star-formation rate (SFR) in the universe (see Porciani & Madau 2001 for a recent review in the context of GRBs). The most massive stars (collapsars) explode soon ($\lesssim 10^7$ yr) after zero-age main sequence (ZAMS) whereas merging neutron stars require a median time to merge of ~ 2–10×10^8 yr since ZAMS (e.g., Phinney 1991; Narayan et al. 1992; Portegies Zwart & Spreeuw 1996; chapter 2, Bloom et al. 1999a). In principle, therefore, due to the significant time from ZAMS to the mergers of NS–NS and BH–NS binaries, such merging remnants should produce GRBs at preferentially *lower* redshift than collapsars and promptly bursting binaries (BH–He); I quantified this "redshift offset" of delayed merging binaries in chapter 2 (see also Fryer et al. 1999a). There are, unfortunately, a number of biases in the observed GRB sample (and universal SFR measurements themselves!) that preclude such a quantitative comparison to solidify the stellar-origin connection and distinguish between progenitors[4]. To date, no one has adequately accounted for all of the observational biases to determine the true, underlying rate; as more uniformly detected redshifts become possible with *Swift*,[5] though, such biases may one day be quantifiable.

More uniformly selected are the associated host galaxies of GRBs and thereby as a sample may be more informative than redshifts. Observer-frame R-band magnitudes of GRB hosts appear, for instance, to be in rough agreement with the hypothesis that the GRB rate follows the SFR of the universe (Mao & Mo 1998; Hogg & Fruchter 1999). The GRB hosts themselves, too, appear to be a fair representation of the luminosity function of the general field population (Djorgovski et al. 2001); that is, GRB hosts do not appear to be extraordinarily bright nor faint. By inspection of the host images in figure 6.2, $\sim$50% of the hosts could be classified as a type irregular, peculiar, or merger, consistent with the fraction from

---

[4] Higher redshift afterglows should be, in general, more dim and then systematically be observed less frequently. Very high redshift bursts ($z \gtrsim 6$) would escape detection at optical wavelengths due to blanketing from the Lyman α forest. Moreover, the window for redshift discovery is not uniform for all redshifts; this is especially true for redshifts discovered by emission spectroscopy of associated host galaxies. So in practice, distinguishing the GRB(z) rate from the SFR(z) rate is exceedingly difficult (e.g., Blain & Natarajan 2000) without tens if not hundreds more GRB redshift measurements (see fig. 2.4).

[5] The *Swift* satellite is a GRB MIDEX mission scheduled for launch in late 2003. The satellite will localize GRBs to sub-arcsecond resolution just seconds after trigger at a rate of about 2–3 per week. The on-board optical telescope will be used for high-quality photometry and spectroscopy of the afterglows. See Barthelmy (2000) and Burrows et al. (2000) for details.

redshift unity galaxies studies (Reshetnikov 2000; Le Fèvre et al. 2000). Furthermore, the distribution of unobscured star-formation, as proxied by the equivalent width of [O II] $\lambda3727$, appears to follow the [O II] distribution of galaxies at $z \sim 1$ (Djorgovski et al. 2001).

The galaxy scale: The ability to compare the gross properties of GRB hosts with other samples is, of course, possible only after studies on the individual hosts themselves. Chapters 3–5 represent some of the first moderately detailed observations of GRB host galaxies. Chapter 5 (on the host of GRB 970228) is the most recent of these studies and shows, I believe, the power of combining detailed spectroscopy with high-resolution imaging to arrive at a rather complete picture of the nature of individual GRB hosts. In that chapter, I strongly refuted a claim (that was based on photometry of the host alone) that the host was extraordinary when compared to other galaxies at similar redshifts (Fruchter et al. 1999).

Before the detailed modeling of light curves were used to constrain the nature of supernovae progenitors, the location of supernovae in and around galaxies provided important clues to the nature of the progenitors (e.g., Reaves 1953; Johnson & MacLeod 1963). For instance, only Type Ia supernovae have been found in elliptical galaxies naturally leading to the idea that the progenitor population can be quite old whereas the progenitors of Type II and Type Ibc are likely to be closely related to recent star formation (see van Dyk 1992, for review). Further, in late-type galaxies, Type Ibc and Type II supernovae appear to be systematically closer to HII star forming regions than Type Ia supernovae (e.g., Bartunov et al. 1994). This is taken as strong evidence that the progenitors of Type Ibc and Type II SNe are massive stars (see Fillipenko 1997).

Similarly, given the delayed time to merge, the instantaneous rate of GRBs from binary mergers is more a function of the integrated (as opposed to instantaneous) star formation rate in its parent galaxy. So if GRBs arise from the death of massive stars we do not expect early-type (i.e., elliptical and S0) host galaxies, whereas GRBs from merging remnants could occur in such galaxies. Indeed, no elliptical host galaxy has yet been uncovered.

More important, independent of galaxy type, the locations of GRBs within (or outside) galaxies provide a measurable signal to help distinguish between progenitor scenarios. Massive stellar explosions occur very near their birth-site, likely in active HII star-forming regions, since the time since ZAMS is so small. BH–He binaries will merge quickly and so are also expected to be located near star-forming regions (Fryer et al. 1999a). In stark contrast, as explored in chapter 2, NS–NS (and NS–BH) binaries merge far from their birthsite. These stellar remnant progenitors will merge after at least one of the binary members has undergone a supernova. Each supernova is thought to impart a substantial "kick" on the resulting neutron star (e.g., Hansen & Phinney 1997); for those binary systems which survive both supernovae explosions, the center-of-mass of the remnant binary itself will receive a velocity boost on the order of a few hundred km s^{-1} (e.g., Brandt & Podsiadlowski 1995). That is, NS–NS or NS–BH binaries will be ejected from their birthsite. The gradual angular momentum loss in the binary due to gravitational radiation causes the binary to coalesce/merge which then leads to a GRB. The time to merge ($\sim 10^6$–10^9 yr) depends on the masses of the remnants and binary orbit parameters. Population synthesis models have all shown that roughly one-third to one-half of NS–NS and BH–NS binary mergers will occur beyond 10 kpc in projection from the centers of their hosts (Bloom et al. 1999a; Fryer et al. 1999a). The distribution of merger sites depends sensitively on the gravitational potential of the host and the (radial) distribution of massive star birth sites.

The realization that the offset of GRBs from their host galaxies could be one of the cleanest observational tests for the progenitors motivated the work in chapter 6. There, by observing the distribution of 20 cosmological GRBs about their hosts, I find good evidence for a progenitor population which follows the UV light of their host galaxies.

1.3.2 An instrument to study the small-scale environments of GRBs

GRBs and afterglows both influence and are influenced by the immediate surroundings. As noted in §1.1, the afterglow stage begins at distance of $\sim 10^{14}$ cm from the explosion site and at such a distance the impact of the progenitor should be felt. For instance, Waxman & Loeb (1999) noted that the mass-loss history from a collapsar progenitor is likely to be complex and non-axisymmetric. The resulting density inhomogeneities would result in small-scale variations of GRB light curves about an otherwise smooth synchrotron shock afterglow. Though the measurement of such features is clearly of interest, after a few hours the r.m.s. fluxuations on timescales of minutes are less than 5%. Practically, this implies that the imprints of the small-scale environments cannot be measured by small aperture telescopes.

To this end, we built the Jacobs Camera instrument (JCAM[6]) for the Hale 200-inch telescope at Palomar Observatory. The instrument—a dual CCD imaging camera with very quick readout—has just ended its commissioning phase and we hope to make the instrument public within a year. Since JCAM is permanently mounted at the East Arm, we can begin observing simultaneously in two optical bandpasses in 5–20 minutes from trigger. I helped to find funding for the instrument, design and integrate the electronics hardware, design and implement the software control systems, integrate the instrument and the telescope, and perform the necessary calibrations and maintenance. I have been solely responsible for the commissioning of JCAM.

Chapter 10 is a version of the JCAM data paper which has been submitted to *PASP* for publication. There, I argue that rapid observations of GRB afterglows can yield important clues to the nature of the progenitors. By mapping the evolution of the afterglow in time and in broadband colors, we hope to be able to directly probe density inhomogeneities (or lack thereof) on scales of 10^{14} cm or smaller and constrain the initial Lorentz factor of GRBs (see §10.2).

Wherefore another optical imaging instrument?

The past decade saw the advent of large-format CCDs and IR arrays for astronomical imaging, fostered by a wide range of scientific objectives (e.g., surveys to detect weak lensing, microlensing events, near-Earth asteroids, planetary transits, or Lyman-break galaxies) and only recently made practical by the rapid fall in detector cost and, most important, data storage. At Palomar Observatory alone three new large-format instruments have emerged as the dominant paradigm for new instrumentation at the site—on the Hale 200 inch, the *Large Format Camera* (LFC[7]) for optical imaging and the *Wide Field Infrared Camera* (WIRC[8]) for infrared imaging; on the Oschin 48 inch Schmidt, the *Near-Earth Asteroid Tracking* system (NEAT[9]). Despite the advantages of a large field, costs and data rates from large-format cameras are still formidable. The LFC, for example, costs over $500 k in hardware alone and, on a typical night generates over 8 GB of data. Given that the mounting and cooling of the prime-focus instruments requires over 24 hrs, such instruments are clearly not well-suited for unanticipated transient follow-up.

On the 200 inch telescope and on the Keck telescopes, the primary instrument tends to change every several days depending on lunar phase and the observers' science. An informal survey of the Keck I 2001B schedule shows that only 45% of the nights were scheduled with an imager suitable for deep optical follow-up. The other prime disadvantages of large-telescope optical instrumentation are the rather large full-frame readout times (e.g., 70 sec for the LFC, 145 sec for COSMIC), large data rates, and the inability to observe many bands simultaneously.

JCAM is an entirely different approach to instrumentation at the Palomar. The scientific intent in the construction of JCAM, described more fully in §10.2, was to provide quick access to multi-color optical imaging of GRB afterglows. The instrument is now continuously mounted at the East Arm f/16 focus of the Hale 200 inch and can be accessed by the installation of the Coudé mirror and secondary mirror in the light path (see §10.4). One advantage of a small number of pixels per image (2.6×10^5) is that data transfer rates, even over T1-line quality connections, are manageable; this has allowed us to observe with JCAM remotely over the Internet, the first such instrument at Palomar Observatory.

---

[6] This instrument was privately funded by a donation from M. Jacobs.
[7] 5×10^7 pix, 0.12 deg^2; http://www.astro.caltech.edu/~ras/lfc/lfc.html
[8] 4×10^6 pix, 0.02 deg^2; http://isc.astro.cornell.edu/~don/wirc.html
[9] 5×10^7 pix, 3.75 deg^2; http://neat.jpl.nasa.gov/neatoschincam.htm

1.3.3 The stellar scale: connection to supernovae

Over the span of just one month in 1998, two GRBs (980326 and 980425) were detected that continue to influence our understanding of GRB progenitors, in particular the connection of GRBs to supernovae and hence to massive stars. GRB 980425 was associated with a nearby peculiar supernovae SN 1998bw at a distance of 39.1 Mpc, suggesting the burst had an extraordinarily low gamma-ray energy output (8×10^{47} erg) compared with the energetics of other GRBs (chapter 9). GRB 980326, as described in chapter 7, was the first cosmological GRB for which an associated supernova was found.

A sub-class of under-luminous GRBs produced by supernovae

In the localization error box of the *BeppoSAX* WFC, Galama et al. (1998b) discovered a Type Ib/Ic supernova which was later recognized as more energetic compared to type Ib/Ic supernovae as measured from optical expansion velocities (Iwamoto et al. 1998). Though, on a purely phenomenological basis, the chance probability for a spurious association between SN 1998bw and GRB 980425 appeared to be small (Galama et al. 1999), there was some ambiguity as to the whether a *bone fide* X-ray afterglow was detected at a position inconsistent with the SN (Pian et al. 1999). Regardless of the true physical connection with GRB 980425, we recognized that the prompt radio emission detected from SN 1998bw necessitated relativistic shock propagation during the initial period of the explosion (Kulkarni et al. 1998a; Kulkarni et al. 1998)—the first evidence for a relativistic shock in a supernova (Wieringa et al. 1999). That the young shock likely contained enough energy ($\gtrsim 10^{49}$ erg) to power a (weak) GRB strengthened the connection between the two phenomena on a *physical* basis. Indeed, it is now widely accepted that the physical association is real (e.g., Wheeler 2001; Salmonson 2001).

Chapter 9 was written against the backdrop of this exciting possibility of a physical connection between a supernova and a gamma-ray burst. There, I realized that the physical model put forth by Kulkarni et al. (1998a) held some predictions about the general properties of the resulting supernovae and accompanying GRB (of course, GRB 980425 and SN 1998bw, naturally exhibited all of these properties). Could this be a newly discovered subclass of GRBs? I asked whether any of the known GRBs and supernovae fit the proposed properties of what I call S-GRBs (supernova GRBs). Despite provocative speculations by a number of other authors about a few other GRB/SN connections, I unfortunately (but not unexpectedly) found no convincing evidence for another plausible association.

Given the lack of a believable association of any other SN/GRB pair, I then sought to place constraints on the frequency of such S-GRBs finding that at most a few percent of such bursts comprise the known GRB sample; figure 9.1 shows some examples of possible S-GRBs based solely on the similarity of the light curves with GRB 980425.

GRB 980326: the first connection of cosmological GRBs to supernovae

GRB 980326 was one the softest and faintest GRBs localized by *BeppoSAX*, but was otherwise unremarkable as a GRB. In the first few days, the afterglow exhibited a rapid decline seen in many other GRBs. As had become standard practice, about 30 days after the burst we imaged the GRB field and obtained a long integration spectrum of the supposed host galaxy. The idea was to find the redshift of the GRB through emission spectroscopy. Though continuum was detected, no obvious lines were seen in the "host" spectrum. Seeking to refine the astrometry so as to improve the blind-offset spectroscopy observations, we re-observed the field of GRB 980326 eight months after the burst and to our surprise the "host" was gone, having faded by at least a factor of 10 in flux from our intermediate-time imaging and spectroscopic detections. Two years later, a faint host was finally detected at the afterglow position using *HST* (see §6.5.5).

A redress of the "host" hypothesis was clearly warranted, leading us to conclude that we had seen something unusual in the month after GRB 980326. Our conclusions (as well as observations) are presented in chapter 7. There I discuss how the data are consistent with the presence of an underlying supernova which peaked in brightness around the same time as our intermediate-time observations. Moreover, I found that the supernova component was consistent with the peak flux of SN 1998bw if the the supernova had occurred at a redshift of unity. A number of alternative physical explanations had been put forth that might have explained the intermediate-time "bump" (e.g., dust heating and re-radiation, delayed energy input from a long lasting remnant, thermal expansion after a merger of a binary neutron star). Yet, as noted in §7.3, all such interpretations failed to explain either the timescale of the bump or the colors of the bump.

Soon after this chapter was submitted, a re-analysis of the optical photometry of GRB 970228 showed the light curve to be consistent with a supernova component (Reichart 1999; Galama et al. 2000). Unfortunately, the flux of the afterglow of GRB 970228 was of a brightness comparable to the SN component, and so the significance of the detection of the bump is greatly diminished relative to that in the GRB 980326. Unlike with GRB 980326, however, GRB 970228 showed photometric evidence for a bump in three bandpasses and showed a roll-over in a broadband spectrum consistent with that of a type Ib/Ic SN (Reichart 2001). Most important, the redshift of GRB 970228 was known (see chapter 5), which removed the peak-brightness—redshift degeneracy that existed in GRB 980326.

GRB 011121: multi-color observations of a supernova-like component

GRB 011121 holds the record as the cosmological GRB with the lowest known redshift ($z = 0.36$). For this reason, the peak in any associated supernova component was expected to be at ~ 7000 Å, squarely in the observer-frame optical bands. Starting about two weeks after the burst, we triggered a series of *HST* observations of the source with the hope of constraining the nature of any intermediate-time emission. As discussed in chapter 8, such emission is seen as a clear excess above that expected from the early afterglow.

Indeed, chapter 8 builds upon the previous emission bump observations in 980326 and 970228, showing excess at four different epochs in as many as five optical filters per epoch. The expected spectral roll-over beyond 7000 Å, if the source of emission was a supernova, is seen. Further, the characteristic rise and decay of core-collapsed supernovae on timescales of weeks, is also seen in the emission bump. I argue that the physical origin of the emission bump is from a supernova which occurred at nearly the same time as the GRB itself.

Part I
The Large-scale Environments of Gamma-Ray Bursts

CHAPTER 2

The Spatial Distribution of Coalescing Neutron Star Binaries: Implications for Gamma-Ray Bursts[†]

JOSHUA S. BLOOM[1] AND STEINN SIGURDSSON[2], AND ONNO R. POLS[2,3]
[1] California Institute of Technology, MS 105-24, Pasadena, CA 91106 USA
[2] Institute of Astronomy, Madingley Road, Cambridge, CB3 0HA, England
[3] Instituto de Astrofísica de Canarias, c/ Vía Láctea s/n, E-38200 La Laguna, Tenerife, Spain

Abstract

We find the distribution of coalescence times, birthrates, spatial velocities, and subsequent radial offsets of coalescing neutron stars (NSs) in various galactic potentials accounting for large asymmetric kicks introduced during a supernovae. The birthrates of bound NS–NS binaries are quite sensitive to the magnitude of the kick velocities but are, nevertheless, similar (~ 10 per Galaxy per Myr) to previous population synthesis studies. The distribution of merger times since zero-age main sequence is, however, relatively insensitive to the choice of kick velocities. With a median merger time of $\sim 10^8$ yr, we find that compact binaries should closely trace the star formation rate in the Universe. In a range of plausible galactic potentials (with $M_{\rm galaxy} \gtrsim 3 \times 10^{10} M_\odot$) the median radial offset of a NS–NS merger is less than 10 kpc. At a redshift of $z = 1$ (with $H_0 = 65$ km s^{-1} Mpc^{-1} and $\Omega = 0.2$), this means that half the coalescences should occur within ~ 1.3 arcsec from the host galaxy. In all but the most shallow potentials, 90 percent of NS–NS binaries merge within 30 kpc of the host. We find that although the spatial distribution of coalescing neutron star binaries is consistent with the close spatial association of known optical afterglows of gamma-ray bursts (GRBs) with faint galaxies, a non-negligible fraction (~ 15 percent) of GRBs should occur well outside ($\gtrsim 30$ kpc) dwarf galaxy hosts. Extinction due to dust in the host, projection of offsets, and a range in interstellar medium densities confound the true distribution of NS–NS mergers around galaxies with an observable set of optical transients/galaxy offsets.

SECTION 2.1

Introduction

The discovery of an X-ray afterglow (Costa et al. 1997b) by BeppoSAX (Costa (Boella et al. 1997) and subsequently an optical transient associated with gamma-ray burst (GRB) 970228 van Paradijs

---
[†] A version of this chapter was first published in *The Monthly Notices of the Royal Astronomical Society*, 305, p. 763–768 (May 1999).

et al. (1997) led to the confirmation of the cosmological nature of GRBs Metzger et al. (1997b). The broadband optical afterglow has been modeled relatively successfully (Mészáros & Rees 1993; Wijers et al. 1997; Waxman 1997a; Waxman et al. 1998) as consistent with an expanding relativistic fireball (Rees & Mészáros 1994; Paczyński & Rhoads 1993; Katz 1994; Mészáros & Rees 1997a; Vietri 1997; Sari et al. 1998; Rees & Mészáros 1998). Still, very little is known about the nature of the progenitors of GRBs, and, for that matter, their hosts. Broadband fluence measures and the known redshifts of some bursts implies a minimum (isotropic) energy budget for GRBs of $\sim 10^{52-53}$ ergs (Metzger et al. 1997b; Kulkarni et al. 1998b; see also Bloom et al. 2001b and fig. 1.2). The log N-log P brightness distribution, the observed rate, N, of bursts above some flux, P, versus flux, indicates a paucity of dim events from that expected in a homogeneous, Euclidean space. With assumptions of a cosmology, source evolution and degree of anisotropy of emission, the log N-log P has been modeled to find a global bursting rate. Assuming the bursts are non-evolving standard candles Fenimore & Bloom (1995) found ~ 1 burst event per galaxy per Myr (GEM) to be consistent with the observed log N-log P. More recently, Wijers et al. (1998) (see also, Totani 1997; Lipunov et al. 1997) found the same data consistent with GRBs as standard candles assuming the bursting rate traces the star-formation rate (SFR) in the Universe; such a distribution implies a local burst rate of ~ 0.001 GEM and a standard peak luminosity of $L_0 = 8.3 \times 10^{51}$ erg s^{-1} (Wijers et al. 1998).

Given the energetics, burst rate and implied fluences, the coalescence, or merger, of two bound neutron stars (NSs) is the leading mechanism whereby gamma-ray bursts are thought to arise (Paczyński 1986; Goodman 1986; Eichler et al. 1989; Narayan et al. 1992). One quantifiable prediction of the NS–NS merger hypothesis is the spatial distribution of GRBs (and GRB afterglow) with respect to their host galaxies. Conventional wisdom, using the relatively long–lived Hulse-Taylor binary pulsar as a model, is that such mergers can occur quite far ($\gtrsim 100$ kpc) outside of a host galaxy. Observed pulsar (PSR) binaries with a NS companion provide the only direct constraints on such populations, but the observations are biased both toward long lived systems, and systems that are close to the Galactic plane.

The merger rate of NS–NS binaries has been discussed both in the context of gravitational wave-detection and GRBs (e.g., Phinney 1991; Narayan et al. 1991; Narayan et al. 1992; Tutukov & Yungelson 1994; Lipunov et al. 1995). Recently Fryer et al. (1998); Lipunov et al. (1997); Portegies Zwart & Spreeuw (1996) studied the effect of asymmetric kicks on birthrates of NS–NS binaries, but did not quantify the spatial distribution of such binaries around their host galaxies. Tutukov & Yungelson (1994) discussed the spatial distribution of NS–NS mergers but neglected asymmetric kicks and the effect of a galactic potential in their simulations. Only Portegies Zwart & Yungelson (1998) have discussed the maximum travel distance of merging neutron stars including asymmetric supernovae kicks.

It is certainly of interest to find the rate of NS–NS coalescences *ab initio* from population synthesis of a stellar population. This provides an estimate of beaming of GRBs, assuming they are due to NS–NS mergers, and hence an estimate of probable frequency of gravitational wave sources, providing a complementary rate estimate to those of Phinney (1991) and Narayan et al. (1991), which are based on long lived NS-PSR pairs only and are very conservative. It also provides an estimate of how the NS–NS mergers trace the cosmological star formation rate (SFR) of the Universe, if mean formation rates and binarity of high mass stars are independent of star formation environments such as metallicity.

Here we concentrate on estimating the spatial distribution of coalescing NS–NS binaries around galaxies. To do so, both the system velocity and the interval between formation of the neutron star binary and the merger through gravitational radiation is found by simulation of binary systems in which two supernovae occur. We explore the effects of different asymmetric kick amplitudes, and the resultant birthrates and spatial distribution of coalescing NS–NS binaries born in different galactic potentials.

In section 2 we briefly outline the prescription for our Monte Carlo code to simulate bound binary pairs from an initial population of binaries by including the effect of asymmetric supernovae kicks. In section 3 we outline the integration method of NS–NS pairs in various galactic potentials. Section 4 highlights the birthrates and spatial distributions inferred from the simulations. Section 5 concludes by discussion the implications and predictions for gamma-ray burst studies.

SECTION 2.2
Neutron Star Binary Population Synthesis

We used a modified version of the code created for binary evolution by Pols (Pols & Marinus 1994) taking into account the evolution of eccentricity through tidal interaction and mass transfer before the first and second supernova, and allowing for an asymmetric kick to both NSs during supernovae. The reader is referred to Pols & Marinus (1994) for a more detailed discussion account of the binary evolution code.

2.2.1 Initial conditions and binary evolution

In general, the evolution of a binary is determined by the initial masses of the two stars (m_1, m_2), the initial semi-major axis (a_o) and the initial eccentricity (e_o) of the binary at zero-age main sequence (ZAMS). We construct Monte Carlo ensembles of high-mass protobinary systems (with primary masses between $4M_\odot$ and $100M_\odot$) by drawing from an initial distribution of each of the four parameters as prescribed and motivated in Portegies Zwart & Verbunt (1996). We treat mass transfer and common-envelope (CE) phases of evolution as in Pols & Marinus (1994). CE evolution is treated as a spiral-in process; we use a value of $\alpha = 1$ for the efficiency parameter of conversion of orbital energy into envelope potential energy; see equation [17] of Pols & Marinus (1994). We treat circularization of an initially eccentric orbit as in Portegies Zwart & Verbunt (1996).

During detached phases of evolution we assume that mass accreted by the companion is negligible so that aM_{tot} = constant. Mass lost by the binary system in each successive time step results in a change in eccentricity according to the sudden mass loss equations (see, for example, eqns. A.21 and A.24 of Wettig & Brown 1996). We ignore the effect of gravitational radiation and magnetic braking in the early stages of binary evolution.

The simple approximation of the 4-parameter distribution function, albeit rather *ad hoc*, appears to adequately reproduce the observed population of lower mass stars in clusters (e.g., Pols & Marinus 1994). The effect on the distribution of NS–NS binaries after the second supernova by variation of the 4-parameter space is certainty of interest, but we have used the canonical values. A fair level of robustness is noted in that varying the limits of the initial distributions of a_o and e_o does not the change the implied birthrates of bound NS binaries nearly as much as plausible variations in the asymmetric kick distribution. This effect was noted in Portegies Zwart & Spreeuw (1996) and Portegies Zwart & Verbunt (1996).

2.2.2 Asymmetric supernovae kicks

Several authors (e.g., Paczyński 1990; Narayan & Ostriker 1990; Lyne & Lorimer 1994; Cordes & Chernoff 1997) have sought to constrain the distribution of an asymmetric kick velocities from observations of isolated pulsars which are the presumed by-products of type II supernovae. Even careful modeling of the selection effects in observing such pulsars has yielded derived mean velocities that differ by nearly an order of magnitude. It is important here to use a good estimate for the actual physical impulse (the "kick velocity") the neutron stars receive on formation. The observed distribution of pulsar velocities does not reflect the kick distribution directly as it includes the Blaauw kick (Blaauw 1961) from those pulsars formed in binaries, and selection effects on observing both the high and low speed tail of the pulsar population (e.g., Hartman et al 1997). Hansen & Phinney (1997) found that the observed distribution is adequately fit by a Maxwellian velocity distribution with $v_{kick} = 190$ km s^{-1} (which corresponds to a 3-D mean velocity of 300 km s^{-1}). Since it is not clear that pulsar observations require a more complicated kick-velocity distribution, we chose to adopt a Maxwellian but vary the value of σ_{kick}.

When a member of the binary undergoes a supernova, we assume the resulting NS receives a velocity kick, v_k, drawn from this distribution. Although the direction of this kick may be coupled to the orientation of the binary plane, we choose a kick with a random spatial direction, since there is no known correlation between the kick direction or magnitude and the binary parameters.

If α is the angle between the velocity kick and the orbital plane and v is relative velocity vector of the two stars, then, following earlier formulae (e.g., Portegies Zwart & Verbunt 1996; Wettig & Brown 1996), the new-semi major axis of the binary is

$$a' = \left(\frac{2}{r} - \frac{v^2 + v_k^2 + 2vv_k\cos\alpha}{G(M_{NS} + M_2)}\right)^{-1} \quad (2.1)$$

where r is the instantaneous distance between the two stars before SN, M_2 is the mass of the companion (which may already be a NS), and $M_{NS} = 1.4 M_\odot$ is the mass of the resulting neutron star. We neglect the effects of supernova-shell accretion on the mass of the companion star. If a' is positive, the new eccentricity is

$$e' = \left[1 - \frac{|\vec{r} \times \vec{v}_r|^2}{a'G(M_{NS} + M_2)}\right]^{1/2}, \quad (2.2)$$

where the resultant relative velocity is $\vec{v}_r = \vec{v} + \vec{v}_k$. Assuming the kick directions between successive SN are independent, the resulting kick to the bound system (whose magnitude is given by equation 2.10 of Brandt & Podsiadlowski 1995) is added in quadrature to the initial system velocity to give the system velocity (v_{sys}).

To produce 1082 bound NS–NS binaries with a Hansen & Phinney kick velocity distribution and initial conditions described above, we follow the evolution of 9.7 million main sequence binaries which produce a total of $\sim$ 1 million neutron stars through supernovae. Assuming a supernova rate of 1 per 40 years (Tammann et al. 1994) and 40% binary fraction (as in Portegies Zwart & Spreeuw 1996), we find an implied birthrate of NS–NS binaries by computing the number of binaries with SN type II per year and multiply by the ratio of bound NS–NS systems to SN type II as found in the simulations. We neglect the (presumed small) contribution of other formation channels (e.g., three-body interactions) to the overall birthrate of NS–NS binaries. The implied birthrate of NS–NS binaries from various kick-velocity magnitudes are given in table 2.

SECTION 2.3
Evolution of Binaries Systems in a Galactic Potential

The large-scale dynamics of stellar objects are dominated by the halo gravitational potential while the initial distribution of stellar objects is characterized by a disk scale length. We take the disk scale and halo scale to vary independently in our galactic models. We assume that the NS–NS binaries are born in an exponential stellar disk, with birthplace drawn randomly from mass distribution of the disk. The initial velocity is the local circular velocity (characterized but the halo) plus v_{sys} added with a random orientation.

We then integrate the motion of the binary in the galactic potential assuming a Hernquist (1990) halo; we ignore the contribution of the disk to the potential. We assume scale lengths for the disk and halo: the disk scale (r_{disk}) determines the disk distribution, the halo scale length (r_{break}) and circular velocity (v_{circ}) determine the halo mass (see table 1). The movement of the NS–NS binaries on long time-scales is sensitive primarily to the depth of the galactic potential (here assumed to be halo dominated) and how quickly it falls off at large radii. Assuming isothermal halos instead of Hernquist profiles would decrease the fraction of NS–NS pairs that move to large galactocentric radii, but the differences in distribution are dominated by the true depth of the halo potentials in which the stars form rather than their density profiles at large radii.

We use a symplectic leapfrog integrator to advance the binary in the galactic potential, and a simple iteration scheme to evolve the semi-major axis and eccentricity of the binary as gravitational radiation drives a and e to zero, assuming the orbit averaged quadrupole dominated approximation (Peters 1964).

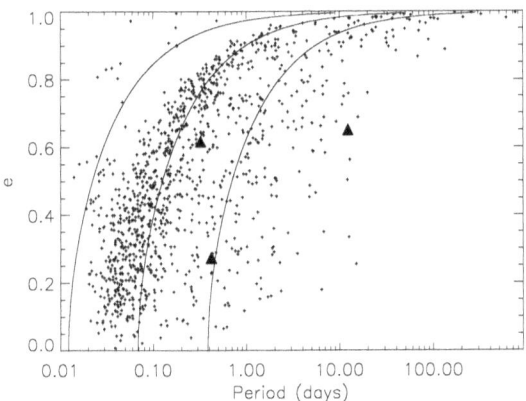

Figure 2.1 The distribution of orbital parameters (period and eccentricity) after the second supernovae for bound NS–NS pairs. From left to right are lines of constant merger time after second SN (10^6, 10^8, 10^{10} yrs). The parameters of the observed NS pairs 1913+16 (Taylor & Weisberg 1989), 1534+12 (Wolszczan 1991), and 2303+46 (Taylor & Dewey 1988) are marked with triangles. With an observational bias toward long-lived systems, clearly the observed PSR-NS systems are not indicative of the true NS binary distribution.

Table 2.1. The Spatial Distribution of Coalescing Neutron Star Binaries in Various Galactic Potentials

Run	Galaxy parameters				L	Coalescence Distance	
	v_{circ} (km s^{-1})	r_{break} (kpc)	r_{disk} (kpc)	M ($10^{11} M_\odot$)		d_{median} (kpc)	d_{avg} (kpc)
a	100	1	1	0.092	$\lesssim 0.05 L_*$	4.3	66.2
b	100	3	1	0.278	$\simeq 0.1 L_*$	4.0	50.1
c	100	3	3	0.278	$\simeq 0.1 L_*$	8.7	68.8
d	150	3	1	0.625	$\simeq 0.5 L_*$	3.1	24.8
e	150	3	3	0.625	$\simeq 0.5 L_*$	7.7	54.1
f	225	3	3	1.41	$\simeq 1 L_*$	6.0	29.9
g	225	3	1	1.41	$\simeq 1 L_*$	2.3	7.1
h	225	5	3	2.34	$\simeq 2 L_*$	6.0	21.4
i	225	5	5	2.34	$\simeq 2 L_*$	9.9	30.2

Note. — Though the average distance from center a pair travels before coalescence (d_{avg}) generally decreases with increasing galactic mass, the median distance (d_{median}) scales with disk radius (r_{disk}).

The integration is continued until either 1.5×10^{10} years have passed (no merger in Hubble time) or the characteristic time to merger is short compared to the dynamical time scale of the binary in the halo (i.e., the binary will not move any further before it merges). We then record the 3-D position of the binary relative to the presumptive parent galaxy and the time since formation.

SECTION 2.4

Results

2.4.1 Orbital parameter distribution after the second supernova

Figure 2.1 shows the distribution of orbital parameters (semi-major axis and period) after the second supernova for bound NS–NS pairs for the Hansen & Phinney (1997) kick distribution ($\sigma_{\mathrm{kick}} = 190$ km s^{-1}). As found previously (e.g., Portegies Zwart & Spreeuw 1996), bound systems tend to follow lines of constant merger time. The density of systems in figure 2.1 can be taken as the probability density of finding a NS–NS binaries directly after the second supernova. In time, the shorter-lived systems (higher e and shorter period) merge due to gravitational radiation. Thus, at any given time after a burst of star-formation there is an observational bias toward finding longer-lived systems. In addition, there is a large observational bias against finding short period binaries (Johnston & Kulkarni 1991). That the observed PSR-NS systems lie in the region of parameter space with low initial probability is explained by these effects. The time-dependent probability evolution has been discussed and quantified in detail by Portegies Zwart & Yungelson (1998). Figure 2.2 shows the distribution of merger times as a function of system velocity. A majority of systems merge in $\sim 10^8$ yr spread over system velocities of 50 – 500 km s^{-1}. A subclass of systems have spatial velocities and merger times that are anti-correlated.

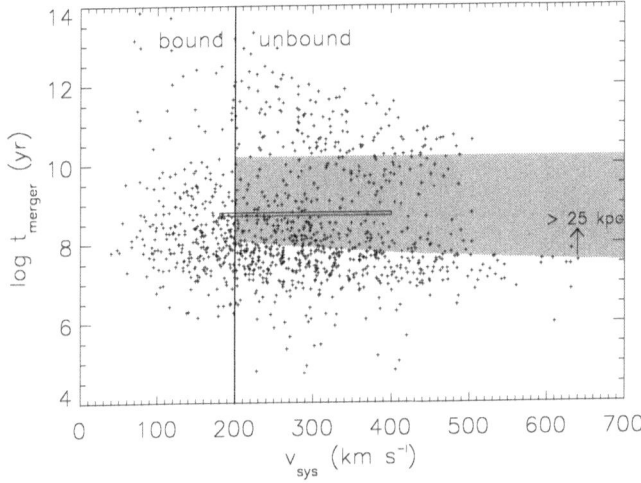

Figure 2.2 The distribution of merger times after second supernovae as a function of system velocity. Left of the vertical line, all pairs created are gravitationally bound to an under-massive host ($3 \times 10^{10} M_\odot$) at the disk scale radius. Of the pairs that are unbound, only the pairs in the shaded region could travel more than ~ 25 kpc (linearly) from their birthplace and merge within a Hubble time ($\lesssim 1.5 \times 10^{10}$ yrs). Since the spatial velocity of observed NS binaries includes both the initial circular velocity of the system and the system velocity due to kicks from each supernova, the true system velocities are highly uncertain. For comparison, though, we denote the range of accepted kick velocities of PSR 1913+16 with a long rectangle (the merger time is much better constrained than that depicted); this illustrates a general agreement of the system velocity of PSR 1913+16 and the modeled distribution of bound NS binaries. The slightly longer merger time of PSR 1913+16 than expected from the density of systems in this parameter space is explained in section 4.1 of the text.

Table 2.2. The Bound NS–NS Binary Birthrate and Merger Time Properties as a Function of Supernova Kick Strength

σ_{kick} (km s^{-1})	Birthrate (Myr $^{-1}$)	τ_{median} (yr)	τ_{avg}^a (yr)
95	49	1.4×10^8	9.4×10^8
190	10	7.0×10^7	8.0×10^8
270	3	5.5×10^7	7.0×10^8

Note. — ^{a}Average merger time of pairs merging in less than 1.5×10^{10} years. A Maxwellian distribution characterized by a velocity dispersion (σ_{kick}) is assumed.

2.4.2 Coalescence/birth rates

We have explored the consequences of different kick strengths on the birthrates of NS–NS binaries. Table 2.2 summarizes these results.

Earlier work (e.g., Sutantyo 1978; Dewey & Cordes 1987; Verbunt et al. 1990; Wijers et al. 1992; Brandt & Podsiadlowski 1995) in which asymmetric kicks were incorporated with a single NS component binary (as in LMXBs, HMXBs) noted a decrease in birthrate with increased kick magnitude. Portegies Zwart & Spreeuw (1996) and Lipunov et al. (1997) found a similar effect on the bound NS pair birthrates. Lipunov (1997) provides a good review of the expected rates. Clearly, the birthrate of NS–NS binaries is also sensitive to the total SN type II rate (which is observationally constrained to no better than a factor of two, and theoretically depends both on the uncertain high mass end of the initial mass function and the total star formation at high redshift), and is also sensitive to the fraction of high mass stars in binaries with high mass secondaries.

We concentrate our discussion of NS–NS binary birthrates to galactic systems for which the SN type II is fairly well-known (such as in the Galaxy). It is important to note, however, that the SN type II rate may be quite high in low surface-brightness and dwarf galaxies (e.g., Babul & Ferguson 1996). This would subsequently lead to a higher NS–NS birthrates in such systems than a simple mass scaling to rates derived for the Galaxy.

Recently van den Heuvel & Lorimer (1996) find (observationally) the birthrate of NS–NS binaries to be 8 Myr^{-1}. Lipunov et al. (1997) find between 100 and 330 events per Myr in simulations. Portegies Zwart & Spreeuw (1996) found birthrates anywhere from 9 to 384 Myr^{-1} depending mostly on the choice of asymmetric kick strength in their models. We note that our derived birthrate of ~ 3 Myr^{-1} for high $\sigma_v = 270$ km s^{-1} is comparable to those found Portegies Zwart & Spreeuw (particularly model "ck") with an average 3-D kick velocity of 450 km s^{-1}. Also, for low velocity kicks ($\sigma_{kick} = 95$ km s^{-1}) our birthrates approach that of Portegies Zwart & Spreeuw models with no asymmetric kicks.

The discrepancies between this and other work, therefore, we believe, are largely due to the choices of supernovae kick distributions and strengths. That the absolute birthrate varies by an order of magnitude depending on the binary evolution code and asymmetric kick distributions used in different studies hints at the uncertainty in the *ab initio* knowledge of the true birthrates.

2.4.3 Spatial distribution

Approximately half of the NS–NS binaries merge within $\sim 10^8$ years after the second SN; this merger time is relatively quick on the timescale of star-formation. In addition, half of the pairs coalesce within

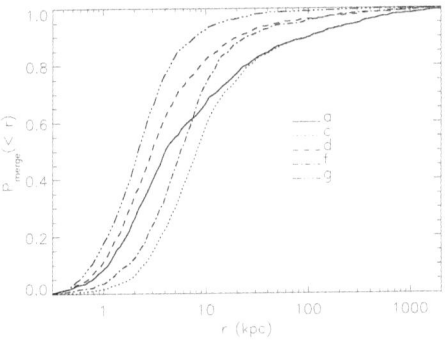

Figure 2.3 The radial distribution of coalescing neutron stars around galaxies of various potentials. The letters refer to runs in table 1. In all scenarios, at least 50% of the mergers occur within 10 kpc of the host galaxy. The wider radial distribution of in the under-luminous galaxy scenarios (a,c) reflects the smaller gravitational potential of under-luminous galaxies.

a few kpc of their birthplace and within 10 kpc of the galactic center (see figure 2.3) *regardless* of the potential strength of the host galaxy. As shown in figure 2.3, galaxies with $M_{galaxy} > 10^{10} M_\odot$ ($L \gtrsim 0.1 L_*$), 90 (95) percent of the NS-NS mergers will occur within 30 (50) kpc of the host. In the least massive dwarf galaxies with $M_{galaxy} \simeq 9 \times 10^9 M_\odot$ ($\lesssim 0.1 L_*$), 50 (90, 95) percent of mergers occur within $\sim$ 10 (100, 300) kpc of the host (see figure 2.3). So, for example, assuming a Hubble constant of $H_0 = 65$ km s^{-1} Mpc^{-1} and $\Omega = 0.2$, we find that 90 (95) percent of NS binaries born in dwarf galaxies at redshift $z = 1$ will merge within $\sim$ 12.7 arcsec ($\sim$ 38.2 arcsec) of the host galaxy. These angular offsets can be considered the extreme of the expected radial distribution since the potentials are weakest and we have not included the effect of projection. We would expect 50 (90, 95) percent of the mergers near non-dwarf galaxies to occur within $\sim$ 1.3 (3.8, 6.4) arcsec from their host at $z = 1$ for the cosmology assumed above.

Given the agreement of our orbital parameter distribution (figure 1) and velocity distribution (figure 2) with that of Portegies Zwart & Yungelson (1998), the discrepancy between the derived spatial distribution (see figure 8 of Portegies Zwart & Yungelson) is likely due to our use of a galactic potential in the model. This inclusion of a potential naturally keeps merging NSs more concentrated toward the galactic center than without the effect.

SECTION 2.5
Discussion

Although the NS-NS birthrate decreases with increased velocity kick, the distribution of merger time and system velocity is not affected strongly by our choice of kick distributions. Rather, the shapes of the orbital and velocity distributions (figures 1 and 2) are closely connected with the pre-SN orbital velocity, which is itself connected simply with the evolution and masses. That is, bound NS binaries come from a range of parameters which give high orbital velocities in the pre-second SN system. The orbital parameters (and merger time distribution) of binaries which survive the second SN are not sensitive to the exact kick-velocity distribution. We suspect this may be because bound systems can only originate from a parameter space where the kick magnitude and orientation are tuned for the

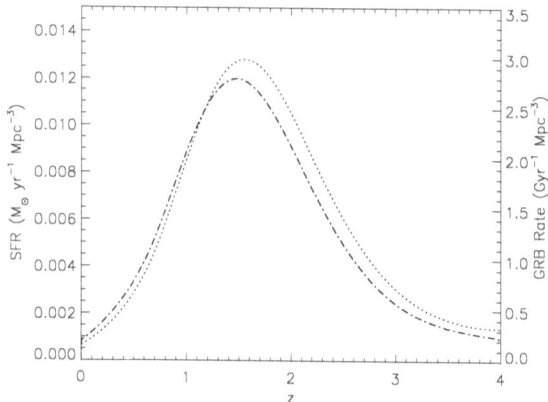

Figure 2.4 NS–NS merger rate dependence on redshift. The dotted line is a reproduction of the SFR from Madau (1997) with corresponding units on the left–hand axis. The SFR curve as seen as a lower limit to the true star-formation history since dust may obscure a large fraction SFR regions in galaxies. The right hand axis is the (unobscured) GRB rate if the bursts arise from the merger of two NS–NS assuming a merger time distribution found in the present study (dot–dashed line). Both the SFR and merger rate are in co-moving units (assuming $H_0 = 50$ km s^{-1} Mpc^{-1}). The normalization of the burst rate is taken from Wijers et al. (1998).

pre-second SN orbital parameters. The overall fraction of systems that remain bound *is* sensitive to the kick distribution insofar as the kick distribution determines how many kicks are in the appropriate range of parameter space.

Since NS–NS binaries are formed rapidly (with an average time since ZAMS of ~ 22 million years) and the median merger time is of order one hundred million years regardless of the kick velocity distribution (see table 2), the rate of NS–NS mergers should closely trace the star formation rate. In the context of gamma-ray bursts, where merging NSs are seen as the canonical production mechanism, this result implies that the GRB merger rate should evolve proportionally to the star formation rate (see figure 2.4; see also Bagot et al. 1998). Indeed, several studies (Totani 1997; Lipunov et al. 1997; Wijers et al. 1998) have consistently fit the GRB log N–log P curve to a model which assumes such a rate density evolution.

If indeed gamma-ray bursts arise from the coalescence of neutron star binaries, then we confirm that GRBs should trace the star formation rate in the Universe; thus most GRBs should have redshifts near the peak in star formation (currently believed to be $1 \lesssim z \lesssim 2$; Madau et al. 1998) although the observed distribution may be skewed to lower redshifts by obscuration at high redshift (e.g., Hughes et al. 1998). Determination of the distribution of X-ray and optical counterparts to GRBs may help constrain the true cosmological star formation history, though the observations of GRB counterparts are vulnerable to some of the same extinction selection effects that complicate determination of high redshift star formation rates. Figure 2.4 illustrates the redshift dependence of the GRB rate assuming the bursts arise in NS–NS mergers.

The minimum required local (isotropic) bursting rate of 0.025 galactic event per Myr (Wijers et al. 1998) is consistent with our birthrate results (table 2) assuming a beaming fraction of $1/10$–$1/100$ for the gamma ray emission and our canonical values for the type II supernova rate and supernova binary fraction. The effects of beaming should be observed in both the light curves of GRB afterglow and in deep transient searches (e.g., Woods & Loeb 1998).

In the case of GRB 970508, Bloom et al. (1998b) (chapter 3) and Castro-Tirado et al. (1998b) (see Natarajan et al. 1997) found that the host is an under-luminous dwarf galaxy; the close spatial connection (offset < 1") of the OT with the galaxy is then a case (albeit weak) against the NS–NS merger hypothesis as the *a priori* probability is $\lesssim 20\%$ (figure 2.3). Paczyński (1998) first pointed out that the close spatial association with a dwarf galaxy is a case against the NS–NS merger hypothesis. Certainly more transients are required to rule against the NS–NS merger hypothesis; we note, however, that dust obscuration and projection effects may severely bias the sample (see above discussion).

The verdict on the expected radial distribution of NS–NS mergers with hosts of known GRBs is still out. Sahu et al. (1997) found the optical transient associated with GRB 970228 to be slightly offset from the center of a dim galaxy, but without a redshift it is still unclear as to the the true luminosity of the host and thus the expected offset of the OT in the NS–NS merger hypothesis. Similarly, small or negligible offsets of GRB afterglows with faint galaxies has been found in other GRBs. Kulkarni et al. (1998b) found the redshift of the purported host galaxy of GRB 971214 to be $z = 3.4$ implying that the host is $L \gtrsim L_*$; the expected offsets of NS–NS mergers around massive galaxies (figure 2.3, models d through i) is then consistent with their finding of an OT offset $\simeq 0.5$ kpc.

A few well-established offsets cannot tell us what is the true distribution of GRBs around host galaxies. As more OTs are discovered, we will hopefully build up a large sample to statistically test the offsets. Fortunately, the unobscured afterglow emission strength is coupled with the density n of the surrounding interstellar medium (ISM) with intensity scaling as $\sqrt{n}$ (Begelman et al. 1993; Mészáros et al. 1998); however, high ISM densities tracing dust will tend to obscure rest-frame UV and optical emission from the transient. In the absence of strong absorption from the surrounding medium, transients of GRBs are preferential found close to where they are born, in the disk. However, dust obscuration and projection effects severely complicate determination of the true offset of OTs from their host galaxy. Furthermore, identification of the host with a GRB becomes increasingly difficult with distances beyond a few light radii (~ 10 kpc) of galaxies.

If all afterglows, especially those where little to no absorption is implied, are found more highly concentrated than predicted in figure 2.3, the NS–NS merger hypothesis would lose favor to models which keep progenitors more central to their host. GRBs as events associated with single massive stars such as microquasars (Paczyński 1998) or failed type Ib SN (Woosley 1993) could be possible. Alternatively, one may consider neutron star–black hole (BH) binaries as the progenitors of GRBs (Mochkovitch et al. 1993). Most black hole X-ray binaries have low-spatial velocities (although Nova Sco has $v_{\text{sys}} \simeq 100$ km s^{-1}; see Brandt et al. 1995) so NS–BH binaries should have system velocities ~ 3 to 10 times smaller than NS–NS binaries. One would expect NS–BH systems to be borne with higher eccentricities than NS–NS systems leading to quicker merger. Moreover, NS–BH binaries are more massive than NS–NS binaries and merger time due to gravitational radiation scales strongly with mass. Thus the attraction is that NS–BH mergers would be preferentially closer to their host and their merger rate might be small enough so as to require no beaming. Alternatively, gamma-ray bursts could arise from several of these plausible progenitor models and still be consistent with basic relativistic fireball models.

SECTION 2.6
Conclusions

A reconciliation with the expected distribution of presumed progenitors of GRBs and observed transient/host offsets is clearly required. We find for all plausible galactic potentials that the median radial offset of a NS–NS merger is less than 10 kpc. And in all but the most shallow potentials, ninety percent of NS–NS binaries merge within 30 kpc of the host. At a redshift of $z = 1$ (with $H_0 = 65$ km s^{-1} Mpc^{-1}

and $\Omega = 0.2$), this means that ninety percent the coalescences should occur within ~ 4 arcsec from the host galaxy. Although the expected spatial distribution of coalescing neutron star binaries found herein is consistent with the close spatial association of known optical afterglows of gamma-ray bursts with faint galaxies, a non-negligible fraction ($\sim 15\%$) of GRBs *should* occur well outside dwarf galaxy hosts if the NS–NS hypothesis is correct. Otherwise, other models which keep progenitors closer to their host (e.g., BH–NS mergers, microquasars, or failed SN type Ib SNe) would be preferred.

As all the progenitor models mentioned are connected with high-mass stars, the true GRB afterglow rate as a function of redshift should trace the star-formation rate in the universe. However, environmental effects, such as dust obscuration, may severely bias the estimate of the true offset distribution. Even in the NS–NS models where progenitors have a natural mechanism to achieve high spatial velocities, most will be closely connected spatially to their host. Redshifts derived from absorption in the afterglow spectra should be nearly always that of the nearest galaxy (Bloom et al. 1997). Rapid burst follow-up ($\lesssim 1$ hr), with spectra taken while the optical transients are bright should confirm some form of absorption from the host galaxy.

We have confirmed the strong dependence of birthrate of NS–NS binaries on kick velocity distribution and found the independence of the orbital parameters after the second supernova (and hence merger times and spatial velocity) on the choice of kicks. The methodology herein can be extended to include formation scenarios of black holes. This could provide improved merger rate estimates for LIGO sources, and estimate the relative contribution of coalescences between neutron stars and low mass black holes to the event rate. Detailed modeling of the Milky Way potential would also allow predictions for the distribution of NS–PSR binaries observable in the Milky Way, which would provide an independent test of the assumptions made in these models.

It is a pleasure to thank Peter Mészáros, Melvyn Davies, Gerald Brown, Hans Bethe, Ralph Wijers, Peter Eggleton, Sterl Phinney, Peter Goldreich, Brad Schaefer, and Martin Rees for helpful insight at various stages of this work. We especially thank Simon Portegies Zwart as referee. JSB thanks the Hershel Smith Harvard Fellowship for funding. SS acknowledges the support of the European Union through a Marie Curie Individual Fellowship.

CHAPTER 3

The Host Galaxy of GRB 970508[†]

J. S. Bloom, S. G. Djorgovski, S. R. Kulkarni

California Institute of Technology, Palomar Observatory, 105-24, Pasadena, CA 91125

D. A. Frail

National Radio Astronomy Observatory, P. O. Box O, Socorro, NM 87801

Abstract

We present late-time imaging and spectroscopic observations of the optical transient (OT) and the host galaxy of GRB 970508. Imaging observations roughly 200 and 300 days after the burst provide unambiguous evidence for the flattening of the light-curve. The spectroscopic observations reveal two persistent features which we identify with [O II] $\lambda\lambda3727$ Å and [Ne III] $\lambda3869$ Å at a redshift of $z = 0.835$ — the same redshift as the absorption system seen when the transient was bright. The OT was coincident with the underlying galaxy to better than 370 milliarcsec or a projected radial separation of less than 2.7 kpc. The luminosity of the [O II] line implies a minimum star formation rate of $\gtrsim 1~M_\odot$ yr^{-1}. In our assumed cosmology, the implied restframe absolute magnitude is $M_B = -18.55$, or $L_B = 0.12 L_*$. This object, the likely host of GRB 970508, can thus be characterized as an actively star-forming dwarf galaxy. The close spatial connection between this dwarf galaxy and the OT requires that at least some fraction of progenitors be not ejected in even the weakest galactic potentials.

SECTION 3.1

Introduction

After an initial brightening lasting ~1.5 days, the optical transient (OT) of GRB 970508 faded with a nearly pure power-law slope by 5 mag over ~100 days (e.g., Galama et al. 1998a; Garcia et al. 1998; Sokolov et al. 1998). Indications of a flattening in the light curve (Pedersen et al. 1998) were confirmed independently by Bloom et al. (1998c), Castro-Tirado et al. (1998a), and Sokolov et al. (1998). Recently, Zharikov et al. (1998) fit the $BVRI$ light curves of the OT + host and found the broadband spectrum of the presumed host galaxy.

The existence of an [O II] emission line at the absorption system redshift (Metzger et al. 1997a) was taken as evidence for an underlying, dim galaxy host. After *Hubble Space Telescope* (*HST*) images revealed the point-source nature of the light, several groups (e.g., Fruchter et al. 1997; Pian et al. 1998; Natarajan et al. 1997) suggested that the source responsible for the [O II] emission must be a very faint

---

[†] A version of this chapter was first published in *The Astrophysical Journal Letters*, 518, p. L1–L4 (1999).

($R > 25$ mag), compact (less than 1″) dwarf galaxy at $z=0.835$ nearly coincident on the sky with the transient. These predictions are largely confirmed in the present study.

In this Letter, we report on the results of deep imaging and spectroscopy of the host galaxy of GRB 970508 obtained at the 10 m Keck II telescope.

SECTION 3.2
Observations and Analysis

Imaging and spectroscopic observations were obtained using the Low Resolution Imaging Spectrograph (LRIS) (Oke et al. 1995) on the 10 m Keck II Telescope on Mauna Kea, Hawaii. The log of the observations is presented in table 3.1. All nights were photometric. The imaging data were reduced in the standard manner.

To follow the light-curve behavior of the OT + host over $\sim$300 days from the time of the burst, we chose to tie the photometric zero point to a previous study (Sokolov et al. 1998) which predicted late-time magnitudes based on early (less than 100 days) power-law behavior in several bandpasses. This photometric tie provides an internally consistent data set for our purposes. Other studies of the light curve include Galama et al. (1998a) and Pian et al. (1998). V. Sokolov (1998, private communication) provided magnitudes of eight "tertiary" field stars ($R = 18.7$–23 mag) as reference, since the four secondary comparison stars (Sokolov et al. 1998) were saturated in all of our images. The zero points were determined through a least-squares fit and have conservative errors of $\sigma_B = 0.05$ and $\sigma_R = 0.01$ mag.

For our spectroscopic observations, we used a 300 lines mm^{-1} grating, which gives a typical resolution of $\approx$15 Å and a wavelength range from approximately 3900 to 8900 Å. The spectroscopic standards G191B2B (Massey et al. 1988) and HD 19445 (Oke & Gunn 1983) were used to flux-calibrate the data of October and November, respectively. Spectra were obtained with the slit position angle at 51° in order to observe both the host galaxy of GRB 970508 and g1 (see Djorgovski et al. 1997b). This angle was always close to the parallactic angle, and the wavelength-dependent slit losses are not important for the discussion below. Internal consistency implied by measurements of independent standards implies an uncertainty of less than 20% in the flux zero-point calibration. Exposures of arc lamps were used for the wavelength calibration, with a resulting r.m.s. uncertainty of about 0.3 Å and possible systematic errors of the same order, due to the instrument flexure.

SECTION 3.3
Results

Table 3.1 gives a summary of the derived magnitudes at the position of the OT and, as a comparison, the extrapolated magnitudes from a pure power-law decay fit by Sokolov et al. (1998). The OT + host is brighter by greater than 0.8 mag in both the B- and R-band, leading to the obvious conclusion that the transient has faded to reveal a constant source. We used a Levenberg-Marquardt χ^2 minimization method to fit a power-law flux (OT) plus constant flux (galaxy) to the B and R light curves using data compiled in Sokolov et al. (1998):

$$f_{\text{total}} = f_0 t^{-\alpha} + f_{\text{gal}}, \quad (3.1)$$

where t is the time since the burst measured in days. The quantities f_0 and f_{gal} are the normalization of the flux of the transient and the persistent flux of the underlying galaxy, respectively. We find $B_{\text{gal}} = 26.77 \pm 0.35$ mag, $R_{\text{gal}} = 25.72 \pm 0.20$ mag, $B_0 = 19.60 \pm 0.04$ mag, $R_0 = 18.79 \pm 0.03$ mag, with values for the decay parameters of $\alpha_B = -1.31 \pm 0.03$ and $\alpha_R = -1.27 \pm 0.02$. Note that the power-law decline did not start until day ~ 1.6, so the normalizations, B_0 and R_0, do not actually correspond to the true flux of the transient on day 1.

To search for any potential offset of the OT and the galaxy, we used an early image of the gamma-ray burst (GRB) field obtained on COSMIC at the Palomar 200 inch telescope on 1997 May 13.6 UT

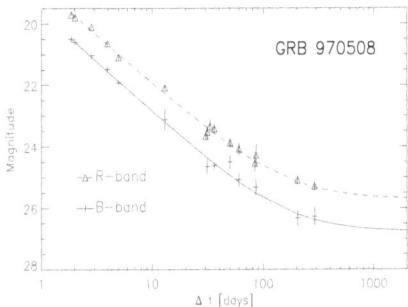

Figure 3.1 Light curve of the optical transient of GRB 970508. Both R- (dashed; triangles) and B-band (solid; crosses) data were compiled and transformed to a single photometric system by Sokolov et al. (1998) (see references therein). The latest two data points on each light curve are from this chapter. The constant flux of the underlying galaxy (the purported host) dominates the light at late times.

while the transient was still bright ($R \approx 20$ mag; see Djorgovski et al. 1997b). Assuming the power-law behavior continued, the light at the transient position is now dominated by the galaxy, with the transient contributing less than 30% to the total flux (see figure 3.1).

We registered the Keck LRIS and the P200 COSMIC R-band (300 s) images by matching 33 relatively bright ($R < 23$ mag) objects in a $4' \times 4'$ field surrounding the GRB transient. The coordinate transformation between the two images accounted for pixel scale, rotating, translation, and higher order distortion. The r.m.s. of the transformed star positions (including both axes) was $\sigma = 0.56$ LRIS pixels ($=0''.121$). We find the angular separation of the OT and the galaxy to be less than 0.814 pixels ($=0''.175$), which includes the error of the transformation and centering errors of the objects themselves. The galaxy is thus found well within 1.7 pixels $= 0''.37$ (3σ) of the OT.

The averaged spectrum of the OT + host shows a very blue continuum, a prominent emission line at $\lambda_{obs} = 6839.7$ Å, and a somewhat weaker line at $\lambda_{obs} = 7097.7$ Å (§3.2). We interpret the emission features as [O II] $\lambda\lambda$ 3727 and [Ne III] λ 3869 at the weighted mean redshift of $z = 0.8349 \pm 0.0003$. Our inferred redshift for the host is consistent (within errors) with that of the absorbing system discovered by Metzger et al. (1997a).

The spectrum of the nearby galaxy g1 shows a relatively featureless, blue continuum. We are unable to determine its redshift at this stage.

Our spectroscopic measurements give a magnitude $R \approx 25.05$ mag (OT + host) at the mean epoch (≈ 163 days after the GRB) of our observations, in excellent agreement with the magnitude inferred from the fit to direct imaging data (see figure 3.1).

Discussion

After an initial brightening, the light curve of the optical transient did not deviate significantly from a power law over the first 100 days after the burst (e.g., Galama et al. 1998a; Garcia et al. 1998; although, see Pedersen et al. 1998). Assuming the blast wave producing the afterglow expanded relativistically (bulk Lorentz factor Γ greater than a few) during the beginning of the light-curve decline, the observed flux was produced from within an angle $\omega_\Gamma \simeq 1/\Gamma$ of the emitting surface. As the blast wave expands,

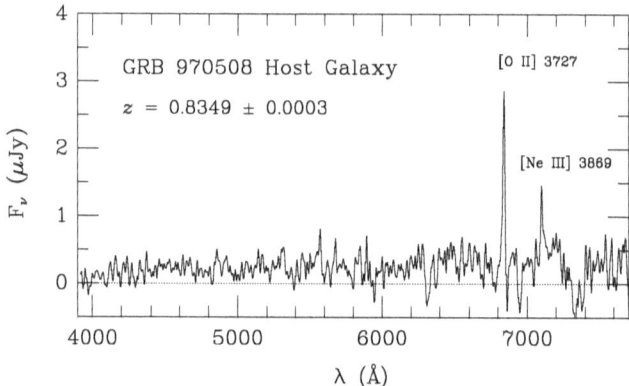

Figure 3.2 The weighted average spectrum of the host galaxy of GRB 970508, obtained at the Keck telescope. The spectrum was smoothed with a Gaussian with a $\sigma = 5$ Å, roughly corresponding to the instrumental resolution. Prominent emission lines are labeled.

ω_Γ increases with time. As long as the angle through which the blast wave is collimated is greater than ω_Γ, there would be no obvious break in the light curve (e.g., Sari et al. 1998). One might expect, in addition, the blast wave to eventually become sub-relativistic, resulting not only in a larger observed surface area but perhaps in a change in surface emissivity. Curiously, an apparent break expected in either scenario did not materialize.

The spatial coincidence of the transient and the underlying galaxy may simply be a chance projection of the transient, which lies beyond $z = 0.835$, and the galaxy at $z = 0.835$. The surface density of galaxies down to $R = 25.7$ mag is 48.3 arcmin^{-2}. It is important to note that we know a priori that the host must lie in the redshift range $0.835 < z < 2.1$ (Metzger et al. 1997b). The fraction of galaxies within this range is ~50% of the total at the magnitude level (Roche et al. 1996). The a posteriori Poisson probability of finding such a galaxy within $0''.37$ from the OT is 3×10^{-3} . Keeping in mind the limitations of a posteriori statistics, this small probability and the trend that GRB transients appear to be nearly spatially coincident with galaxies (e.g., Odewahn et al. 1998) lead us to suggest that this galaxy is the host of GRB 970508.

Assuming a standard Friedman model cosmology with $H_0 = 65$ km s^{-1} Mpc^{-1} and $\Omega_0 = 0.2$, we derive a luminosity distance of 1.60×10^{28} cm to the host galaxy. The observed equivalent width in the [O II] line is (115 ± 5) Å, or about 63 Å in the galaxy's rest frame. However, this also includes the continuum light from the OT at this epoch. Correcting for the OT contribution would then double these values of the equivalent width. This is at the high end of the distribution for the typical field galaxies at comparable magnitudes and redshifts (Hogg et al. 1998). The implied [O II] line luminosity, corrected for Galactic extinction, is $L_{3727} = (9.6 \pm 0.7) \times 10^{40}$ ergs s^{-1}. Using the relation from Kennicut (1998), we estimate the star formation rate (SFR) ≈ 1.4 M$_\odot$ yr^{-1}.

An alternative estimate of the SFR can be obtained from the continuum luminosity at $\lambda_{\rm rest} = 2800$ Å (Madau et al. 1998). The observed, interpolated continuum flux from the host itself (i.e., not including the OT light) at the corresponding $\lambda_{\rm obs} \approx 5130$ Å is $F_\nu \approx 0.11 \mu$Jy, corrected for the estimated Galactic extinction ($A_V \approx 0.08$ mag; Djorgovski et al. 1997b). For our assumed cosmology, the restframe continuum luminosity is then $L_{2800} \approx 1.93 \times 10^{27}$ ergs s^{-1} Hz^{-1}, corresponding to SFR ≈ 0.25 $M_\odot$ yr^{-1}. This is notably lower than the SFR inferred from the [O II] line. We note, however, that neither is known to be a very reliable SFR indicator. Both are also subject to the unknown extinction corrections from the galaxy's own interstellar medium (the continuum estimate being more sensitive). We thus conclude that the *lower limit* to the SFR in this galaxy is probably about 0.5–1 $M_\odot$ yr^{-1}.

The observed flux in the [Ne III] $\lambda 3869$ line is $F_{3869} = (1.25 \pm 0.1) \times 10^{-17}$ erg s^{-1} cm^{-2}, not corrected for the extinction. The flux ratio of the two emission lines is $F_{3869}/F_{3727} = 0.44 \pm 0.05$. This ratio is about 10 times higher than the typical values for H II regions. Nonetheless, it is in the range of photo-ionization models for H II regions by Stasinska (1990) for different combinations of model parameters but generally for effective temperatures $T_{\text{eff}} \geq 40,000$ K.

The inferred host luminosity is in agreement with the upper limit from earlier *HST* observations (Pian et al. 1998). Further, our derived B- and R-magnitudes for the galaxy correspond to a continuum with a power law $F_\nu \sim \nu^{-1.56}$. Extrapolating from the observed R-band flux to the wavelength corresponding to the rest-frame B-band (about 8060 Å), we derive the observed flux F_ν (λ = 8060 Å) $\approx 0.22\mu$Jy. For our assumed cosmology, the implied rest-frame absolute magnitude is then $M_B \approx -18.55$. Thus, the rest-frame B-band luminosity of the host galaxy is about $0.12\, L_*$ today.

This galaxy is roughly 2 mag fainter than the knee of the observed luminosity function of all galaxies between redshift z= 0.77 and 1.0 (Canada-France redshift survey; Lilly et al. 1995) and 1 mag fainter than late-time, star-forming galaxies in the 2dF survey (Folkes et al. 1999). The specific SFR per unit luminosity is high. This object can thus be characterized as an actively star-forming dwarf galaxy. Objects of this type are fairly common at comparable redshifts.

SECTION 3.5

Conclusions

The high effective temperature implied by the relative line strengths of [Ne III] and [O II] suggests the presence of a substantial population of massive stars and thus active and recent star formation. This, in turn, gives additional support to the ideas that the origin of GRBs is related to massive stars (e.g., Wijers et al. 1998; Totani 1997; Djorgovski et al. 1998). An alternative possibility for the origin of the [Ne III] $\lambda 3869$ line is photo-ionization by a low-luminosity active galactic nucleus (AGN). While we cannot exclude this possibility, we note that there is no other evidence in favor of this hypothesis, and moreover we see no other emission lines, e.g., Mg II $\lambda 2799$, that would be expected with comparable strengths in an AGN-powered object.

What may be surprising, in the neutron star binary (NS–NS) model of GRB progenitors (e.g., Narayan et al. 1992; Paczyński 1986), is that GRB 970508 appears so close (less than 2.7 kpc) to a dwarf galaxy ($L \approx 0.1 L_*$). Bloom et al. (1999a) (chapter 2) recently found that less than 15% of NS-NS binaries will merge within 3 kpc of a comparable under-massive galaxy. If GRBs are consistently found very near (less than a few kpc) their purported host, then progenitor models such as microquasars (Paczyński 1998), "failed" Type Ib supernovae (Woosley 1993), or black hole–neutron star binaries (Mochkovitch et al. 1993; Mészáros & Rees 1997b), all of which are expected to produce GRBs more tightly bound to their hosts, would be favored.

It is a pleasure to thank S. Odewahn, M. van Kerkwijk, R. Gal, and A. Ramaprakrash for assistance during observing runs at Keck, P. Groot for comments, and R. Sari for helpful discussions concerning inferences from the light-curve. SRK's research is supported by the National Science Foundation and NASA. SGD acknowledges a partial support from the Bressler Foundation.

Table 3.1. Late-time GRB 970508 Imaging and Spectroscopic Observations

Date (UT)	Band/Grating	Integration time (sec)	Seeing (arcsec)	Δt (days)	Object Magnitude Observed[a]	Object Magnitude Pure Power Law[b]	Observers
Imaging							
Nov. 28, 1997	R	5400	1.2	203.8	25.09±0.14[c]	25.63±0.2	Kulkarni, van Kerkwijk, and Bloom
Nov. 29, 1997	R	600	1.2	204.8			Kulkarni, van Kerkwijk, and Bloom
Nov. 29, 1997	R	600	1.1	204.8			Kulkarni, van Kerkwijk, and Bloom
Nov. 29, 1997	B	2400	1.1	204.8	26.32±0.26	26.65±0.25	Kulkarni, van Kerkwijk, and Bloom
Feb. 22, 1998	R	2400	1.1	290.5	25.29±0.16	26.07±0.20	Kulkarni, Ramaprakash, and van Kerkwijk
Feb. 23, 1998	B	2400	1.2	291.5	26.27±0.28	27.11±0.25	
Spectroscopy							
Oct. 3, 1997		5400	0.8	147.8			Djorgovski and Odewahn
Nov. 2, 1997		5400	< 1.0	177.8			Djorgovski and Gal

[a]Magnitudes are derived from V. Sokolov's tertiary reference stars. The (conservative) 1 σ errors include statistical uncertainties in the reference transformation and the OT + host itself. All nights (imaging observations) were photometric.

[b]Predicted magnitudes are derived from a pure power-law decline using light curve data from $\Delta t \lesssim 100$ days (Sokolov et al. 1998). Errors are estimated using the uncertainties in both the magnitude scaling and power-law index.

[c]The R-band magnitude quoted is derived from the sum of R-band images over two nights.

CHAPTER 4
The Host Galaxy of GRB 990123[†]

J. S. Bloom[1], S. C. Odewahn[1], S. G. Djorgovski[1], S. R. Kulkarni[1], F. A. Harrison[1], C. Koresko[1], G. Neugebauer[1], L. Armus[2], D. A. Frail[3], R. R. Gal[1], R. Sari[1], G. Squires[1], G. Illingworth[4], D. Kelson[5], F. Chaffee[6], R. Goodrich[6], M. Feroci[7], E. Costa[7], L. Piro[7], F. Frontera[8,9], S. Mao[10], C. Akerlof[11], T. A. McKay[11]

[1] California Institute of Technology, Palomar Observatory, 105-24, Pasadena, CA 91125
[2] Infrared Processing and Analysis Center, Caltech, MS 100-22, 770 S. Wilson Avenue, Pasadena, CA 91125.
[3] National Radio Astronomy Observatory, P.O. Box O, 1003 Lopezville Road, Socorro, NM 87801.
[4] University of California Observatories/Lick Observatory, University of California at Santa Cruz, Santa Cruz, CA 95064.
[5] Department of Terrestrial Magnetism, Carnegie Institute of Washington, 5241 Broad Branch Road, NW, Washington, DC 20015-1305.
[6] W. M. Keck Observatory, 65-0120 Mamalahoa Highway, Kamuela, HI 96743.
[7] Instituto di Astrofisica Spaziale CNR, via Fosso del Cavaliere, Roma, I-00133, Italy.
[8] ITESRE-CNR, Via Gobetti 101, Bologna, I-40129, Italy.
[9] Physics Department, University of Ferrara, Via Paradiso 12, Ferrara, I-44100, Italy.
[10] Max-Planck-Institut fur Astrophysik, Karl-Schwarzschild-Strasse 1, Postfach 1523, Garching, 85740, Germany.
[11] Department of Physics, University of Michigan at Ann Arbor, 2071 Randall, Ann Arbor, MI 48109-1120.

Abstract

We present deep images of the field of GRB 990123 obtained in a broadband UV/visible bandpass with the *Hubble Space Telescope* (*HST*) and deep near-infrared images obtained with the Keck I 10 m telescope. The *HST* image reveals that the optical transient (OT) is offset by $0''.67$ (5.8 kpc in projection) from an extended, apparently interacting galaxy. This galaxy, which we conclude is the host galaxy of GRB 990123, is the most likely source of the absorption lines of metals at a redshift of $z = 1.6$ seen in the spectrum of the OT. With magnitudes of Gunn-$r = 24.5 \pm 0.2$ mag and $K = 22.1 \pm 0.3$ mag, this corresponds to an L $\sim 0.5 L_*$ galaxy, assuming that it is located at $z = 1.6$. The estimated unobscured star formation rate is ≈ 4 $M_\odot$ yr^{-1}, which is typical for normal galaxies at comparable redshifts. There is no evidence for strong gravitational lensing magnification of this burst, and some alternative explanation for its remarkable energetics (such as beaming) may therefore be required. The observed offset of the OT from the nominal host center, the absence of broad absorption lines in the afterglow spectrum, and the relatively blue continuum of the host do not support the notion that gamma-ray bursts (GRBs) originate from active galactic nuclei or massive black holes. Rather, the data are consistent with models of GRBs that involve the death and/or merger of massive stars.

[†] A version of this chapter was first published in *The Astrophysical Journal Letters*, 507, p. L25–L28 (1998).

Section 4.1

Introduction

Following the detection of GRB 990123 by BeppoSAX (Piro et al. 1999b), we discovered an optical transient (OT) (Odewahn et al. 1999) and subsequently a coincident radio transient (Frail & Kulkarni 1999) within the error circles of the gamma-ray burst (GRB) and the associated fading X-ray source (Piro et al. 1999a). Examination of the ROTSE (Robotic Optical Transient Search Experiment) images taken during the GRB itself revealed a hitherto unseen bright ($m \simeq 8.9$ mag at peak) phase of the optical afterglow (Akerlof et al. 1999).

In Kulkarni et al. (1999a) we present a comprehensive study of the optical and infrared observations of the transient afterglow and report a measurement of an absorption redshift of $z_{abs} = 1.6$. Combining the redshift with the observed fluence (Feroci et al. 1999) results in an inferred energy release of 3×10^{54} ergs (if the emission was isotropic), which clearly poses a problem to most conventional models of GRBs. However, noting a break in the optical afterglow decay, Kulkarni et al. (1999a) argue that the emission geometry may have been jet-like; this would then decrease the energy constraint.

Both currently favored models of GRB progenitors, the death of a massive star (Woosley 1993; Paczyński 1998) and the coalescence of a neutron star (NS) or NS–black hole (BH) binary (Paczyński 1986; Goodman 1986; Narayan et al. 1992), predict that GRB rates should correlate strongly with the cosmic star formation rate (SFR), and so most GRBs should occur during epochs of the highest SFR (i.e., a redshift range of $z = 1$–3). The former model predicts a tight spatial correlation between GRBs and star-forming regions in galactic disks. The latter, however, allows the coalescence site of a merging binary component to be quite distant (beyond a few kiloparsecs) from the stellar birth site (e.g., Bloom et al. 1999a; chapter 2). GRBs could also be associated with nuclear black holes (i.e., active galactic nuclei) (e.g., Carter 1992). In this scenario, unlike either model described above, the GRBs will preferentially occur in the center of the host. The exquisite angular resolution of the *Hubble Space Telescope* (*HST*) is well suited to address this issue of the locations of GRBs relative to the host galaxies. In this Letter, we report on HST observations of the host galaxy of GRB 990123 taken about 16 days after the burst as well as Keck imaging in the near-infrared.

Section 4.2

Observations and Data Reduction

The ground-based near-IR images of the field of GRB 990123 were obtained using the near-infrared camera Matthews & Soifer (1994) on the Keck I 10 m telescope. A log of the observations and a detailed description of the data and the reduction procedures are given by Kulkarni et al. (1999a). The observations were obtained in the K or K_s bands and were calibrated to the standard K band (effective wavelength = 2.195 μm). The Galactic extinction correction is negligible in the K band (see below).

The first evidence of the underlying galaxy, approximately $0''.6$ from the OT, was seen in our Keck K-band images taken on 1999 January 27 UT. The galaxy, the putative host (which we designate as "A"), was then clearly detected in the images obtained on 1999 January 29 UT (Djorgovski et al. 1999c) and later, under excellent seeing conditions, on February 9 and 10 (see figure 4.1). We find the OT and host fluxes as follows. Sets of pixels dominated by the OT or by the galaxy were masked, and total fluxes with such censored data were evaluated in photometric apertures of varying radii. Total fluxes of the OT + galaxy were also measured in the same apertures using the uncensored data. We also varied the aperture radii, and the position and the size of the sky measurement annulus. On February 9 (February 10) UT, we found that the OT contributes 65% (57%) ($\pm$10%) of the total OT + galaxy light. The estimated errors of the fractional contributions of the OT to the total light reflect the scatter obtained from variations in the parameters of these image decompositions. In both epochs, the fractional contribution of the host implies that a flux of the host galaxy is 0.9 ± 0.3 μJy ($K_{host} = 22.1 \pm 0.3$ mag). We assume 636 Jy for the flux zero point of the K band for $K = 0$ mag (Bessell & Brett 1988).

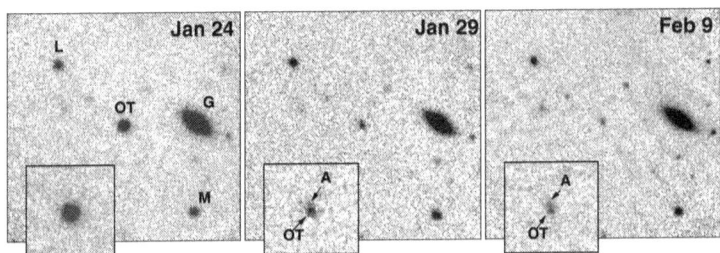

Figure 4.1 Three epochs of Keck I K-band imaging of the field of GRB 990123 (1999 January 24.6, January 29.7, and February 9.6 UT). The field shown is $32'' \times 32''$, corresponding to about 270×270 physical kpc^2 in projection at $z = 1.6004$ (for $H_0 = 65$ km s^{-1} Mpc^{-1} and $\Omega_m = 0.2$, $\Omega_\Lambda = 0.2$). The images have been rotated to the standard orientation, so that the east is to the left and north is up. The magnitude of the host galaxy is $K_{\mathrm{host}} = 22.1 \pm 0.3$ mag. In the January 24 image, the OT dominates the host galaxy flux ($K_{\mathrm{OT}} = 18.29 \pm 0.04$; Kulkarni et al. 1999a), but by January 29 the galaxy is resolved (see inset) from the OT.

The HST observations of the GRB 990123 field were obtained in 1999 February 8–9 UT in response to the director's discretionary time proposal GO-8394, with the immediate data release to the general community (Beckwith 1999). The CCD camera of the Space Telescope Imaging Spectrograph (STIS) (Kimble et al. 1998) in CLEAR aperture (50CCD) mode was used. Over the course of three orbits, the field was imaged in six positions dithered in a spiral pattern for a total integration time of 7200 s. Each position was imaged twice to facilitate cosmic-ray removal (a total of 12 integrations).

Initial data processing followed the STScI pipeline procedures, including bias and dark current subtraction. The six cosmic-ray-removed images were then combined by registering the images and median-stacking to produce a master science-grade image. We also produced a higher resolution image using the "drizzle" technique (Fruchter et al. 1997). Photometry and astrometry were performed on both final image products. We find a negligible difference between the two images compared with other uncertainties (e.g., sky determination and counting noise) in photometry and astrometry. The mean epoch of the final images is 1999 February 9.052 UT.

Figure 4.2 shows a portion of the STIS image of the GRB 990123 field. We find (see below) the OT clearly detected as a point source $0''.67 \pm 0''.02$ to the southwest of the central region of galaxy A. Galaxy A has an elongated and clumpy appearance, possibly indicative of star formation regions in a late-type galaxy. A morphological classification as an irregular galaxy or interacting galaxy system is likely most apt. Such morphologies are typical for many galaxies at comparable flux levels, as observed with the HST. Its extension to the south clearly overlaps with the OT, and it is thus very likely that this galaxy is responsible for the absorption-line system at $z_{\mathrm{abs}} = 1.6004$ (Kulkarni et al. 1999a; Hjorth et al. 1999). The knot "B" to the east may be a satellite of the host galaxy or a star-forming region along the interface zone of a galaxy interacting with the host.

We measured the centroid of the optical transient in our discovery image from January 23 at the Palomar 60 inch telescope (P60). The OT was bright (r = 18.65 mag) at this early epoch, and its position is well determined with respect to other objects in the field. Next we computed the astrometric mapping of the P60 coordinate system to a deep Keck R-band image from 1999 February 9.6 UT (Kulkarni et al. 1999a) using 75 well-detected objects common to the two images. Similarly, we tied the Keck II coordinates to the STIS image using 19 common tie objects. We found the ground-based position of the OT to be consistent with the STIS point source, with a negligible offset of $0''.09 \pm 0''.18$. [Note: after this chapter was published, this source faded in HST imaging, confirming the astrometry.]

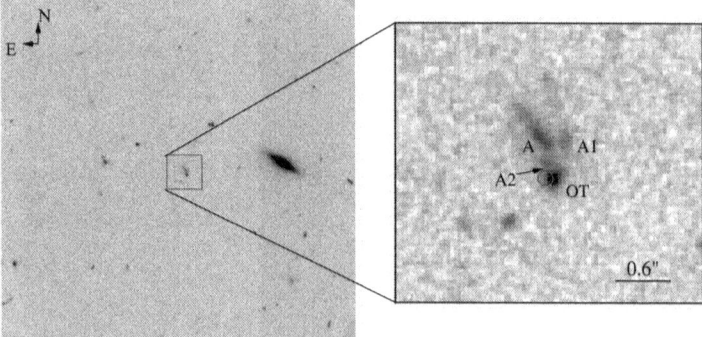

Figure 4.2 The *HST*/STIS "drizzle" image (mean epoch 1999 February 9.052 UT) of the field of GRB 990123 rotated to the normal orientation. (Right) The field shown is 32×32 arcsec^2, corresponding to about 270×270 proper sq. kpc (710×710 co-moving sq. kpc) in projection at $z = 1.6$ (for $H_0 = 65$ km s^{-1} Mpc^{-1} and $\Omega_m = 0.2$). The effective exposure time is 7200 s, and the pixel scale is $0''.0254$ pixel^{-1}. (Left) The OT, galaxy A (the putative host), and the bright knots (A1 and B) associated with galaxy A are denoted. The cross that overlays the OT point source depicts the 1 s uncertainty in each axis for the position of the OT as measured in ground-based imaging (see text). The positional consistency definitively establishes that the point source is indeed the OT. The nebulosity (A2) just to the north of the OT may be a star-forming region (see also Holland & Hjorth 1999).

See §6.5.12.]

The coordinates of the OT as measured in the *HST* image are $\alpha = 15^h25^m30^s.3026$, $\delta = +44^d45'59''.048$ (J2000); the *HST* plate solution is based on the revised guide star catalog and is accurate (in an absolute sense) to $\sim 0''.3$. We note the excellent agreement between this *HST* measurement and the absolute astrometric measurement from ground-based imaging as reported in Kulkarni et al. (1999a). The brightest central region of galaxy A (itself extended north-south) is located $0''.17$ east, $0''.64$ north of the transient. The bright, possibly star-forming, regions "A1" and "B" are located $0''.25$ west, $0''.46$ north and $0''.66$ east, $0''.51$ south, respectively. The uncertainties in the relative positions are $\sim$20 mas.

No other galaxies brighter than $V \sim 27$ mag are detected in the STIS image closer to the OT than galaxy A, and we see no evidence for a distant cluster (or even a sizable group) in this field. This effectively removes the possibility (Djorgovski et al. 1999b) that the burst was significantly magnified by gravitational lensing.

We will assume for the Galactic reddening in this direction $E_{B-V} = 0.016$ mag (Schlegel et al. 1998), and we will use the standard Galactic extinction curve with $R_V = A_V/E_{B-V} = 3.1$ to estimate extinction corrections at other wavelengths. We assume the photometric flux zero points as tabulated by Fukugita et al. (1995).

Most objects detected in the STIS imaging are also detected in the deep Keck II image. After photometric transformation of the R-band Keck II image to the Gunn-r system, we used objects common to both images in order to find a zero point for the STIS image. The r.m.s. uncertainty of the zero point is 0.1 mag, which is mostly due to varied color terms. This scatter implies a transformation of the broadband STIS magnitudes to the Gunn-r system that is rather robust. Taking the system as a whole, $r_{OT+host} = 24.1 \pm 0.1$ mag, which is in excellent agreement with the Keck II imaging (Kulkarni et al. 1999a) taken 12 hr before the *HST* imaging. Aperture photometry on the individual components yields $r_{OT} = 25.3 \pm 0.2$ mag and $r_{host} = 24.5 \pm 0.2$ mag ($F_{\nu,r,host} = 0.5 \pm 0.1$ μJy). The errors are dominated by uncertainties in the color term of the object and the sky value for aperture photometry. Since the host galaxy must contribute some flux in the OT aperture, the magnitude presented above should be considered an upper (lower) limit for the host (OT).

We also determined the OT and host magnitudes by converting observed counts to flux given the instrumental response. Since the bandpass of the STIS CLEAR is so broad, the conversion depends on the assumed spectrum of the object. For the OT, we assumed a spectral index of $\beta \approx -0.8$ (with $F_\nu \propto \nu^\beta$), as inferred from theory and other OTs. For the host, we take $\beta = -0.5$ as a good approximation for star-forming galaxies in the redshift range of the host and also consistent with the K-band measurement. In both cases, we explored a range of plausible spectral indices. Using the STIS exposure simulator available from STScI to convert the observed counts to flux, we find $V_{OT} = 25.4$ and $V_{host} = 24.6$ mag. In a recent paper, Fruchter et al. (1999b), from analysis of the same HST data, find $V_{OT} = 25.45 \pm 0.15$ and $V_{host} = 24.20 \pm 0.15$ mag. The determination of the host magnitude is subject to two systematic uncertainties: the assumed galaxy spectrum and the aperture employed in photometry (see above). Within these errors, these measurements are in agreement. Given the assumed spectral shapes, we also find $r_{OT} = 25.1$ mag and $r_{host} = 24.3$ mag, in agreement with our direct photometric tie to ground-based imaging.

We note that the simple power-law approximation to the broadband spectrum of the galaxy, as defined by our STIS and K-band measurements, is $\beta_{host} \approx -0.5$. This relatively blue color is suggestive of active star formation, but it cannot be used to estimate the SFR directly.

SECTION 4.3

Discussion

In what follows, we assume a standard Friedman model cosmology with $H_0 = 65$ km s^{-1} Mpc^{-1}, $\Omega_m = 0.2$, and $\Omega_\Lambda = 0$. For $z = 1.6004$, the luminosity distance is 3.7×10^{28} cm, and $1''$ corresponds to 8.64 proper kpc or 22.45 co-moving kpc in projection. These values are not appreciably different for other reasonable cosmological world models. For example, with $\Omega_m = 0.3$ and $\Omega_\Lambda = 0.7$ the luminosity distance is 4.1×10^{28} cm, and $1''$ corresponds to 9.12 proper kpc.

We consider it likely that the absorption system at $z_{\text{abs}} = 1.6004$ originates from galaxy A since no other viable candidate is seen in the HST images. The proximity of the center of galaxy A to the OT line of sight ($0''.67 \pm 0''.02$), corresponding to 5.8 proper kpc at this redshift, strongly suggests that the two are physically related. We thus propose that galaxy A is the host galaxy of the GRB. Visual inspection of figure 4.2 suggests that a probability of chance superposition at this magnitude level is very small (see §6.3).

In order to estimate the rest-frame luminosity of galaxy A, we interpolate between the observed STIS and K-band data points using a power law, to estimate the observed flux at $\lambda_{\text{obs}} \approx 11570$ Å, corresponding approximately to the effective wavelength of the rest-frame B band. We obtain $F_{\nu, B, \text{rest}} \approx 0.7$ µJy, corresponding to the absolute magnitude $M_B = -20.0$. Locally, an L_* galaxy has $M_B \approx -20.75$ mag. We thus conclude that this object has the rest-frame luminosity that is $L_{\text{host}} \approx 0.5\, L_{*,\text{local}}$. Given the uncertainty of the possible evolutionary histories, it may evolve to become either a normal spiral galaxy or a borderline dwarf galaxy.

We can make a rough estimate of the SFR from the continuum luminosity at $\lambda_{\text{rest}} = 2800$ Å, following Madau et al. (1998). Using the $F_{\nu, 2800}$ estimates given above, the corresponding monochromatic rest-frame power is $P_{\nu, 2800} = 2.9 \times 10^{28}$ ergs s^{-1} Hz^{-1} (for $\beta = 0$, since it may be appropriate in the UV continuum itself) or $P_{\nu, 2800} = 3.6 \times 10^{28}$ ergs s^{-1} Hz^{-1} (for $\beta = -0.8$). The corresponding estimated unobscured SFRs are ≈ 3.6 and ≈ 4.6 $M_\odot$ yr^{-1}, probably accurate to within 50% or better. This modest value is typical for normal galaxies at such redshifts. It is of course a lower limit, since it does not include any extinction corrections in the galaxy itself or any fully obscured star formation.

Further insight into the physical properties of this galaxy comes from its absorption spectrum, presented in Kulkarni et al. (1999a). The lines are unusually strong, placing this absorber in the top 10% of all Mg II absorbers detected in complete surveys (e.g., Steidel & Sargent 1992). Unfortunately, without a direct measurement of the hydrogen column density, it is impossible to estimate the metallicity of the gas. We note that strong metal-line absorbers are frequently associated with high hydrogen column density systems, such as damped Ly α absorbers. The small scatter of redshift in the individual lines implies a very small velocity dispersion, less than 60 km s^{-1} in the galaxy's rest frame. This implies that the absorber is associated with either a dwarf galaxy or a dynamically cold disk of a more massive system.

The OT is well offset (5.8 kpc) from the central region of the host; this clearly casts doubt on an active galactic nucleus origin of GRBs. However, if the host is indeed an interacting galaxy system, then it is plausible that a massive black hole could be created off-center (as recently suggested by Fruchter et al. 1999b), and thus the position of the GRB 990123 could still permit massive black holes as the progenitors of GRBs. Yet, around such massive BHs, the expectation is that the high-velocity—enshrouding material would give rise to absorption in the GRB afterglow. The clear absence of broad absorption lines in our optical afterglow spectrum, then, does not bode well for the massive black hole hypothesis for the origin of GRBs.

The spatially resolved imaging using HST provides the clearest picture of the relation of GRBs to their hosts. The transient of GRB 970228 is displaced from its host center (Sahu et al. 1997; Fruchter et al. 1999a) but still lies within the half-light radius. GRB 970508, on the other hand, is coincident with the nucleus of its host galaxy to $0''.01$ (Fruchter & Pian 1998; Bloom et al. 1998b). As shown in this Letter, GRB 990123 is separated from the central region of the host and appears to be spatially coincident with a bright star-forming region ("A2"; see figure 4.2). Indeed, Holland & Hjorth (1999) have recently corroborated this claim by noting that the size and luminosity of A2 befit the properties of a generic star-forming region. [Note: later HST imaging did not show the purported region A2. It was probably an artifact of an imperfect PSF determination in the Holland & Hjorth (1999) work.] The close connection of GRBs to their hosts can be extended to results from ground-based astrometry by using the host galaxy magnitude as an objective measure (Odewahn et al. 1996a) of the host size. Using the total magnitude of all known host galaxies to date, we note that the optical transients (except for GRB 990123) lie well within the effective half-light radius.

Copious star formation always appears to be spatially concentrated: along spiral arms, in bright compact H II regions in dwarf irregular galaxies, and in the interface zone of interacting galaxies. Given the morphology of the host (figure 4.2), we suggest that GRB 990123 arose from a star-forming complex in the interface zone of what appears to be a pair of interacting galaxies. This is the first clear case of a GRB associated with an interaction region. Another possible case is the host of GRB 980613. Until now, the GRB star formation connection has been primarily through gross star formation rates obtained from spectroscopic indicators. It is possible that with increasingly larger samples of host galaxies, in analogy to supernovae, the relationship of GRBs to the morphology of the hosts may provide complementary insight into the progenitors of GRBs.

We are grateful to S. Beckwith of STScI for the allocation of the director's discretionary time for this project and to the entire BeppoSAX team and the staff of W. M. Keck Observatory for their efforts. We also thank L. Ferrarese for aiding us with HST observing and the anonymous referee for helpful and clarifying comments. This work was supported in part by a grant from STScI, grants from the NSF and NASA, and the Bressler Foundation.

CHAPTER 5

The redshift and the ordinary host galaxy of GRB 970228[†]

J. S. BLOOM, S. G. DJORGOVSKI, S. R. KULKARNI

Palomar Observatory 105-24, California Institute of Technology, Pasadena, CA 91125, USA;
jsb,george,srk@astro.caltech.edu

Abstract

The gamma-ray burst of 1997 February 28 (GRB 970228) ushered in the discovery of the afterglow phenomenon. Despite intense study of the nearby galaxy, however, the nature of this galaxy and the distance to the burst eluded the community. Here we present the measurement of the redshift of the galaxy, the putative host galaxy of GRB 970228, and, based on its spectroscopic and photometric properties, identify the galaxy as a sub-luminous, but otherwise normal galaxy at redshift $z = 0.695$ undergoing a modest level of star formation. At this redshift, the GRB released an isotropic-equivalent energy of $(1.4 \pm 0.3) \times 10^{52}$ erg (20–2000 keV rest frame). We find no evidence that the host is significantly bluer or is forming stars more vigorously than the general field population. In fact, by all accounts in our analysis (color–magnitude, magnitude–radius, star-formation rate, Balmer-break amplitude) the host properties appear typical for faint blue field galaxies at comparable redshifts.

SECTION 5.1
Introduction

The gamma-ray burst of 1997 February 28 (hereafter GRB 970228) was a watershed event, especially at optical wavelengths. The afterglow phenomenon, long-lived multi-wavelength emission, was discovered following GRB 970228 at X-ray (Costa et al. 1997a) and optical (van Paradijs et al. 1997) wavelengths. Despite intense observations, no radio transient of GRB 970228 was found (Frail et al. 1998); the first radio afterglow (Frail et al. 1997) had to await the next *BeppoSAX* localization of GRB 970508. The basic predictions of the synchrotron shock model for GRB afterglow appeared confirmed by GRB 970228 (e.g., Wijers et al. 1997).

Despite intense efforts, early spectroscopy of the afterglow of GRB 970228 (e.g., Tonry et al. 1997; Kulkarni et al. 1997) failed to reveal the redshift. Spectroscopy of the afterglow of GRB 970508 proved more successful (Metzger et al. 1997b), revealing through absorption lines that the GRB originated from a redshift $z \geq 0.835$. Later spectroscopy and imaging revealed a faint galaxy with $z = 0.835$ at the same location of the afterglow (Bloom et al. 1998b). Through the preponderance of subsequent

---
[†] A version of this chapter was first published in *The Astrophysical Journal*, 554, p. 678–683 (2001).

5.1. INTRODUCTION

redshift determinations (currently 19 in total) and the association of GRBs with faint galaxies, it is now widely believed that the majority of all of long duration ($T \gtrsim 1$ s) gamma-ray bursts originate from cosmological distances and are individually associated with faint galaxies.

Even without a redshift, observations of the afterglow of GRB 970228 in relation to its immediate environment began to shed light on the nature of the progenitors of gamma-ray bursts. Ground-based observations of the afterglow revealed a near coincidence of the GRB with the optical light of a faint galaxy (Metzger et al. 1997c; van Paradijs et al. 1997). Later, deeper ground-based images (Djorgovski et al. 1997a) and high resolution images from the *Hubble Space Telescope* (*HST*) showed the light from the fading transient clearly embedded in a faint galaxy (Sahu et al. 1997) with a discernible offset between the host galaxy centroid and the afterglow. Given the low (albeit *a posteriori*) probability of chance superposition of the afterglow with a random field galaxy (a few percent), this galaxy is presumably the host of the GRB (van Paradijs et al. 1997). This connection is supported by noting that almost all well-localized GRBs, as a class, appear to be statistically connected to a nearby galaxy (Bloom et al. 2002). Throughout this paper we will assume that the galaxy is indeed the host galaxy of GRB 970228. Though by no means definitive, the apparent offset of GRB 970228 from its host rendered an active galactic nucleus origin unlikely (Sahu et al. 1997).

The two most popular progenitor scenarios—coalescence of binary compact stellar remnants and the explosion of a massive star ("collapsar")—imply that the gamma-ray burst rate should closely follow the massive star formation rate in the universe. In both formation scenarios a black hole is created as a by-product; however, the scenarios differ in two important respects. First, only very massive progenitors ($M_{ZAMS} \gtrsim 40\,M_\odot$; Fryer, Woosley, & Hartmann 1999) will produce GRBs in the single star model whereas the progenitors of neutron star–neutron star binaries need only originate with $M_{ZAMS} \gtrsim 8\,M_\odot$. Second, the scenarios predict a distinct distribution of physical offsets (Paczyński 1998; Bloom et al. 1999a) in that the coalescence site of merging remnants could occur far from the binary birthplace (owning to substantial systemic velocities acquired during neutron star formation through supernovae), whereas exploding massive stars will naturally occur in star-forming regions. In relation to the predicted offset of GRB 970228, Bloom et al. (1999a) further noted the importance of redshift to determine the luminosity (and infer mass) of the host galaxy: massive galaxies more readily retain binary remnant progenitors. Thus the relationship of GRBs to their hosts is most effectively exploited with redshift by setting the physical scale of any observed angular offset and critically constraining the mass (as proxied by host luminosity).

The redshift of GRB 970228 also plays a critical role in the emerging supernova–GRB link. Following the report of an apparent supernova (SN) component in the afterglow of GRB 980326 (Bloom et al. 1999c), the afterglow light curves of GRB 970228 were reanalyzed, and both Reichart (1999) and Galama et al. (2000) found evidence for a SN component by way of a red "bump" in the light curve about 1 month after the burst. This interpretation, however, relies critically on the knowledge of the redshift to GRB 970228 to set the rest-frame wavelength of the apparent broadband turnover of the SN component in the observed I band. A lower redshift would, for instance, help make the dust-reradiation model of Waxman & Draine (2000) more viable since the peak of thermal dust emission could be no bluer than 1 μm. On the other hand, a red bump from a high-redshift ($z \gtrsim 1$) GRB is difficult to observe from a SN component since line-blanketing of the UV portion of SNe spectra essentially suppresses the (observer frame) optical flux. Instead, the late-time bumps in high-z GRBs may be more readily explained as dust echoes from the afterglow Esin & Blandford (2000).

Finally, knowledge of the redshift is essential in order to derive the physical parameters of the GRB itself, primarily the energy scale. We now know, for instance, that the typical GRB releases about 10^{52} ergs in gamma rays. The distribution of observed isotropic-equivalent GRB energies is, however, very broad (see Kulkarni et al. 2000).

Recognizing these needs, we implemented an aggressive spectroscopy campaign on the presumed host of GRB 970228, as detailed in §5.2. The redshift determination of the host (and, by assumption, the GRB itself) was first reported by Djorgovski et al. (1999a) and is described in more detail in §5.3.

We then use this redshift and the spectrum of the host galaxy in §5.4 to set the physical scale of the observables: energetics, star formation rates, and offsets. Based on this and photometric imaging from HST, we demonstrate in §5.5 that the host is a sub-luminous, but otherwise normal galaxy.

SECTION 5.2
Observations and Reductions

A finding chart and the coordinate location of the host galaxy of GRB 970228 are given in van Paradijs et al. (1997). Spectra of the host galaxy were obtained on the W. M. Keck Observatory 10 m telescope (Keck II) atop Mauna Kea, Hawaii. Observations were conducted over the course of several observing runs: UT 1997 August 13, UT 1997 September 14, UT 1997 November 1 and 28–30, and UT 1998 February 21–24. The observing conditions were variable, from marginal (patchy/thin cirrus or mediocre seeing) to excellent, and on some nights no significant detection of the host was made; such data were excluded from the subsequent analysis. On most nights, multiple exposures (two to five) of 1800 s were obtained, with the object dithered on the spectrograph slit by several arcseconds between the exposures. The net total useful on-target exposure was approximately 11 hr from all of the runs combined.

All data were obtained using the Low-Resolution Imaging Spectrometer (LRIS; Oke et al. 1995) with 300 lines mm^{-1} grating and a 1.0 arcsec wide long slit, giving an effective instrumental resolution FWHM $\approx$ 12 Å. Slit position angle was always set to 87°, with star S1 (van Paradijs et al. 1997) always placed on the slit, and used to determine the spectrum trace along the chip; galaxy spectra were then extracted at a position 2.8 arcsec east of star S1. Efforts were made to observe the target at hour angles so as to make this slit position angle as close to parallactic as possible. Exposures of an internal flat-field lamp and arc lamps were obtained at comparable telescope pointings immediately following the target observations. Exposures of standard stars from Oke & Gunn (1983) and Massey et al. (1988) were obtained and used to measure the instrument response curve, although on some nights the flux zero points were unreliable owing to non-photometric conditions.

Wavelength solutions were obtained from arc lamps in the standard manner, and then a second-order correction was determined from the wavelengths of isolated strong night-sky lines, and applied to the wavelength solutions. This procedure largely eliminates systematic errors owing to the instrument flexure, and is necessary in order to combine the data obtained during separate nights. The final wavelength calibrations have an r.m.s. $\sim 0.2 - 0.5$ Å, as determined from the scatter of the night-sky line centers. All spectra were then re-binned to a common wavelength scale with a sampling of 2.5 Å (the original pixel scale is ~ 2.45 Å), using a Gaussian with a $\sigma = 2.5$ Å as the interpolating/weighting function. This is effectively a very conservative smoothing of the spectrum, since the actual instrumental resolution corresponds to $\sigma \approx 5$ Å.

Individual spectra were extracted and combined using a statistical weighting based on the signal-to-noise ratio determined from the data themselves (rather than by the exposure time). Since some of the spectra were obtained in non-photometric conditions, the final spectrum flux zero-point calibration is also unreliable, but the spectrum shape should be unaffected. We use direct photometry of the galaxy to correct this zero-point error (see below).

Our uncorrected spectrum gives a spectroscopic magnitude $V \approx 26.3$ mag for the galaxy. Direct photometry from the *HST* imaging data indicates $V = 25.75 \pm 0.3$ (Galama et al. 2000) for the host. Given that some of our spectra were obtained through thin cirrus, this discrepancy is not surprising. Thus, in order to bring our measurements to a consistent system, we multiply our flux values by a constant factor of 1.66, but we thus also inherit the systematic zero-point error of $\sim 30\%$ from the *HST* photometry.

There is some uncertainty regarding the value of the foreground extinction in this direction (see discussion in §5.4.3). We apply a Galactic-extinction correction by assuming $E_{B-V} = 0.234$ mag from Schlegel et al. (1998). We assume $R_V = A_V/E_{B-V} = 3.2$, and the Galactic extinction curve from Cardelli et al. (1988) to correct the spectrum. All fluxes and luminosities quoted below incorporate both the flux zero-point and Galactic-extinction corrections.

SECTION 5.3
The Redshift of GRB 970228

The final combined spectrum of the galaxy is shown in figure 5.1. Two strong emission lines are seen, [O II] 3727 and [O III] 5007, thus confirming the initial redshift interpretation based on the [O II] 3727 line alone Djorgovski et al. (1999a). Unfortunately, the instrumental resolution was too coarse to resolve the [O II] 3727 doublet. The weighted mean redshift is $z = 0.6950 \pm 0.0003$. A possible weak emission line of [Ne III] 3869 is also seen. Unfortunately, the strong night-sky OH lines preclude the measurements of the Hβ 4861 and [O III] 4959 lines, as well as the higher Balmer lines.

The corrected [O II] 3727 line flux is $(2.2 \pm 0.1) \times 10^{-17}$ erg cm^{-2} s^{-1} Hz^{-1}, and its observed equivalent width is $W_\lambda = 51 \pm 4$ Å, i.e., 30 ± 2.4 Å in the restframe. This is not unusual for field galaxies in this redshift range (Hogg et al. 1998). The [Ne III] 3869 line, if real, has a flux of at most 10% of the [O II] 3727 line, which is reasonable for an actively star forming galaxy. The corrected [O III] 5007 line flux is $(1.55 \pm 0.12) \times 10^{-17}$ erg cm^{-2} s^{-1} Hz^{-1}, and its observed equivalent width is $W_\lambda = 30 \pm 2$ Å, i.e., 17.7 ± 1.2 Å in the restframe. For the Hβ line, we derive an upper limit of less than 3.4×10^{-18} erg cm^{-2} s^{-1} Hz^{-1} ($\sim 1\ \sigma$), and $W_\lambda < 7$ Å for its observed equivalent width. We note, however, that this measurement may be severely affected by the poor night-sky subtraction.

The continuum flux at $\lambda_{\rm obs} = 4746$ Å, corresponding to $\lambda_{\rm rest} = 2800$ Å, is $F_\nu = 0.29$ μJy, with a statistical measurement uncertainty of $\sim 10\%$ and a systematic uncertainty of 30% inherited from the overall flux zero-point uncertainty. The continuum flux at $\lambda_{\rm obs} \sim 7525$ Å, corresponding to the restframe B band, is $F_\nu = 0.77$ μJy, with a statistical measurement uncertainty of $\sim 7\%$ (plus 30% systematic).

SECTION 5.4
Implications of the Redshift

For the following discussion, we will assume a flat cosmology as suggested by recent results (e.g., de Bernardis et al. 2000) with $H_0 = 65$ km s^{-1} Mpc^{-1}, $\Omega_M = 0.3$, and $\Lambda_0 = 0.7$. For $z = 0.695$, the luminosity distance is 1.40×10^{28} cm, and 1 arcsec corresponds to 7.65 proper kpc or 13.0 co-moving kpc in projection. By virtue of the close spatial connection of GRB 970228 with the putative host galaxy (see §1), we assume that GRB itself occurred at a redshift $z = 0.695$.

5.4.1 Burst energetics

The gamma-ray fluence (integrated flux over time) is converted from count rates under the assumption of a GRB spectrum, the spectral evolution, and the true duration of the GRB. These quantities are estimated from the GRB data itself but can lead to large uncertainties (a factor of few) in the fluence determination. In Bloom et al. (2001b) we developed a methodology to account for these uncertainties as well as "k-correct" each fluence measurement to a standard co-moving bandpass. Utilizing the redshift reported herein we found that the isotropic-equivalent energy release in GRB 970228 was $(1.4 \pm 0.3) \times 10^{52}$ ergs in the co-moving bandpass 20–2000 keV and a bolometric energy release of $(2.7 \pm 1.0) \times 10^{52}$ ergs (Bloom et al. 2001b).

Since at least some GRBs are now believed to be jetted (e.g., Frail et al. 2001), the true energy release may have been significantly less than that implied if the energy release was isotropic. The degree of "jettedness" in GRBs is most readily determined by the observation of an achromatic break in the afterglow light curves. As such, the slow decline and absence of a strong break in the optical light curve of GRB 970228 (e.g., Galama et al. 1997) suggest that the GRB emission was nearly isotropic (see Sari et al. 1999) and so the knowledge of the total energy release is primarily limited by the accuracy of the fluence measurement.

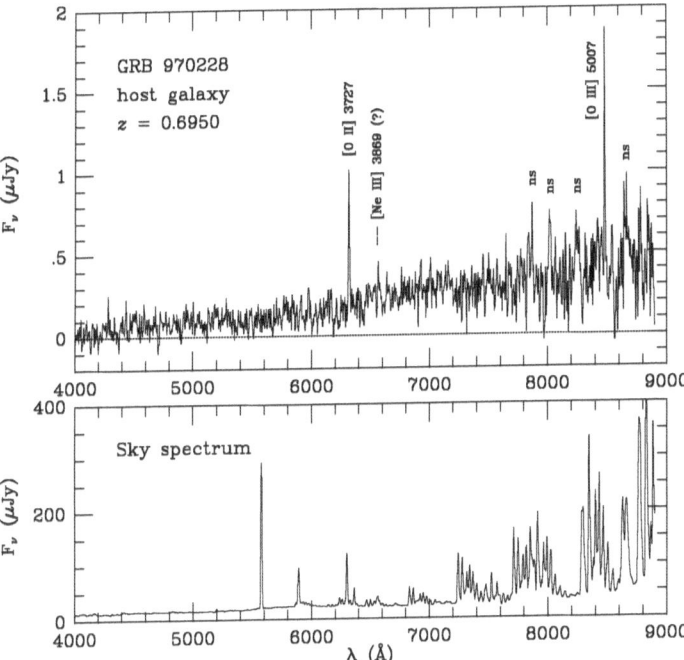

Figure 5.1 *Top*: The weighted average spectrum of the host galaxy of GRB 970228, obtained at the Keck II telescope. Prominent emission lines [O II] 3727 and [O III] 5007 and possibly [Ne III] 3869 are labeled assuming the lines originate from the host at redshift $z = 0.695$. The notation "ns" refers to noise spikes from strong night-sky lines. *Bottom*: The average night-sky spectrum observed during the GRB 970228 host observations, extracted and averaged in exactly the same way as the host galaxy spectrum.

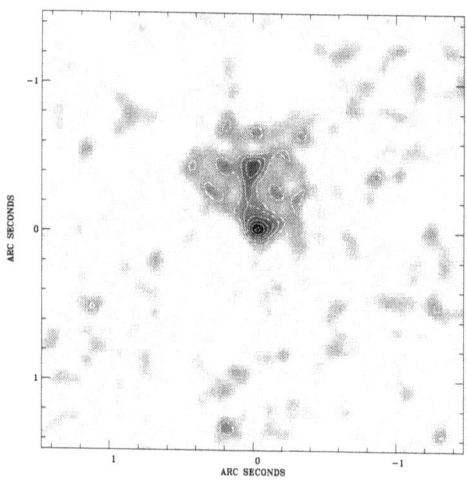

Figure 5.2 A $3'' \times 3''$ (23×23 kpc^2 in projection) region of the HST/STIS image (1997 September 4.7 UT) of the host galaxy of GRB 970228. The image has been smoothed (see text) and is centered on the optical transient. North is up and east is left. Contours in units of 3,4,5,6,7, and 8 background σ ($\sigma = 2.41$ DN) are overlaid. The transient is found on the outskirts of detectable emission from a faint, low surface brightness galaxy. The morphology is clearly not that of a classical Hubble type, though there appears to be a nucleus and an extended structure to the north of the transient.

5.4.2 The offset of the gamma-ray burst and the host morphology

For the purpose of determining the position of the GRB within its host, we examined the HST/STIS observations taken on 1997 Sept 4.7 UT (Fruchter et al. 1999a). The observation consisted of eight 575 s STIS clear (CCD50) exposures paired into four 1150 s images to facilitate removal of cosmic rays. We processed these images using the drizzle technique of Fruchter & Hook (1997) to create a final image with a plate scale of 0.0254 arcsec pixel^{-1}. To enhance the low surface brightness host galaxy, we smoothed this image with a Gaussian of $\sigma = 0.043$ arcsec. The optical transient is well detected in figure 5.2 (point source toward the south) and is clearly offset from the bulk of the detectable emission of the host.

Two morphological features of the host stand out: a bright knot manifested as a sharp 6 σ peak near the centroid of the host pointing north of the transient and an extension from this knot toward the transient. Although, as we demonstrate below, this host is a sub-luminous galaxy (i.e., not a classic late-type L_* spiral galaxy) and we attribute these features to a nucleus and a spiral-arm, respectively. It is not unusual for dwarf galaxies to exhibit these canonical Hubble-diagram structures (S. Odewahn, private communication).

Centroiding the transient and the nucleus components within a 3 pixel aperture radius about their respective peak, we find an angular offset of 436 ± 14 milliarcsec between the nucleus and the optical transient. With our assumed cosmology, this amounts to a projected physical distance of $3.34 \pm 0.11\, h_{65}^{-1}$

kpc.

5.4.3 Physical parameters of the presumed host galaxy

We found the half-light radius of the host galaxy using our final drizzled *HST*/STIS image: we mask a 3×3 pixel region around the position of the optical transient and inspect the curve of growth centered on the central bright knot, the supposed nucleus, and we estimate the half-light radius to be 0.31 arcsec or 2.4 h_{65}^{-1} kpc (physical) at a redshift of $z = 0.695$. The half-light radii of the host galaxy in the *HST*/WFPC F814W (*I*-band) and F606W (*V*-band) filters estimated by eye from the curve-of-growth plots of Castander & Lamb (1999b) are comparable to the STIS-derived half-light radius.

Although there is some debate (at the 0.3 mag level in A_V) as to the proper level of Galactic extinction toward GRB 970228 (Castander & Lamb 1999a; González et al. 1999; Fruchter et al. 1999a), we have chosen to adopt the value $E(B - V) = 0.234$ found from the dust maps of Schlegel et al. (1998) and a Galactic reddening curve $R_V = A_V/E(B - V) = 3.2$. Using extensive reanalysis of the *HST* imaging data by Galama et al. (2000), the extinction-corrected broadband colors of the host galaxy are $V = 25.0 \pm 0.2$ mag, $R_c = 24.6 \pm 0.2$ mag, $I_c = 24.2 \pm 0.2$ mag. These measurements, consistent with those of Castander & Lamb (1999b) and Fruchter et al. (1999a), are derived from the WFPC2 colors and broadband STIS flux. The errors reflect both the statistical error and the uncertainty in the spectral energy distribution of the host galaxy. We have not included a contribution from the uncertainty in the Galactic extinction. Using the NICMOS measurement from Fruchter et al. (1999a), the extinction-corrected infrared magnitude is $H_{AB} = 24.6 \pm 0.1$. Using the zero-points from Fukugita et al. (1996), the extinction-corrected AB magnitudes of the host galaxy are $V_{AB} = 25.0$, $R_{AB} = 24.8$, $I_{AB} = 24.7$.

To facilitate comparison with moderate-redshift galaxy surveys (§5.5), we compute the rest-frame *B*-band magnitude of the host galaxy. From the observed continuum in the rest-frame *B* band, we derive the absolute magnitude $M_B = -18.4 \pm 0.4$ mag [or $M_{AB}(B) = -18.6 \pm 0.4$ mag], i.e., only slightly brighter than the Large Magellanic Cloud (LMC) now. For our chosen value of H_0, an L_* galaxy at $z \sim 0$ has $M_B \approx -20.9$ mag, and thus the host at the observed epoch has $L \sim 0.1~L_*$ today.

5.4.4 Star formation in the host

From the [O II] 3727 line flux, we derive the line luminosity $L_{3727} = 5.44 \times 10^{40}$ erg s^{-1} ($\pm 5\%$ random; $\pm 30\%$ systematic). Using the star formation rate (SFR) estimator from Kennicut (1998), we derive that SFR $\approx 0.76~M_\odot$ yr^{-1}. Using a 3 σ limit on the Hβ flux, we estimate $L_{H\beta} < 2.5 \times 10^{40}$ erg s^{-1}. Assuming the Hα/Hβ ratio of 2.85 ± 0.2 for the Case B recombination and a range of excitation temperatures, we can derive a pseudo Hα-based estimate of the star formation rate (see Kennicut 1998), SFR $< 0.6~M_\odot$ yr^{-1}, but we consider this to be less reliable than the [O II] 3727 measurement. From the UV-continuum luminosity at $\lambda_{\rm rest} = 2800$ Å, following Madau et al. (1998), we derive SFR $\approx 0.54~M_\odot$ yr^{-1}.

We note that the net uncertainties for each of these independent SFR estimates are at least 50%, and the overall agreement is encouraging. While we do not know the effective extinction corrections in the host galaxy itself, these are likely to be modest given its blue colors (see §5.5) and are unlikely to change our results by more than a factor of two. (We hasten to point out that we are completely insensitive to any fully obscured star formation component, if any is present.) On the whole, the galaxy appears to have a rather modest (unobscured) star formation rate, $\sim 0.5 - 1~M_\odot$ yr^{-1}. Given the relatively normal equivalent width of the [O II] 3727 line, even the star formation per unit mass does not seem to be extraordinarily high.

Section 5.5
The Nature of the Host Galaxy

At $M_{AB}(B) = -18.6$ mag, the presumed host galaxy of GRB 970228 is a sub-luminous galaxy, roughly 2.7 mag below L_* at comparable redshifts (Lilly et al. 1995). Based on the redshift of GRB 970228, Galama et al. (2000) recently found that an Sc galaxy spectral energy distribution reasonably fits the optical-IR photometric fluxes of the host galaxy. This differs from the analysis of Castander & Lamb (1999b) who modeled the predicted galaxy colors as a function of the (then unknown) redshift and morphological classification. Now, given the redshift of $z = 0.695$, the Castander & Lamb (1999b) analysis would tend to favor classification as an "Irregular" galaxy, having undergone a burst of star formation over the past few hundred Myr. Clearly it is difficult to precisely determine the galaxy type without more precise photometry and knowledge of the true Galactic extinction, but our identification of a nucleus and possible arm structure (§5.4.2) supports the idea that the host is a late-type dwarf. Indeed, the host has similar characteristics to that of the LMC. We further note that compared to the results of Simard et al. (1999), the magnitude-size relation of the host galaxy is consistent with that observed for late-type and dwarf-irregular galaxies.

The flat continuum suggests a blue continuum of the host galaxy and little rest-frame extinction. How does the host galaxy color compare to other galaxies at comparable magnitudes? We found the $(H - V)_{AB}$ color of galaxies in the Hubble Deep Field-North (HDF-N) using the published photometry from Thompson et al. (1999) (NICMOS F160W filter) and Williams et al. (1996) (WFPC2 F606W filter). All WFPC object identifications within 0.3 arcsec of a NICMOS identification are plotted in figure 5.3 along with the dereddened color of the host galaxy of GRB 970228. This selection from the HDF chooses only those faint objects with detectable IR emission; galaxies with detected visual emission in the WFPC filter but no IR emission in the NICMOS filter do not make it in to our comparison sample. To be clear, this selection essentially biases the color–magnitude relation toward more *red* objects and would serve to accentuate the locus of the comparison field with a blue galaxy. Even with this bias, there is no indication in figure 5.3 that the host galaxy of GRB 970228 is substantially more blue than other field galaxies at comparable magnitudes.

This conclusion—that the host galaxy of GRB 970228 is not exceptionally blue—is at odds with that of Fruchter et al. (1999a) who have claimed that the host galaxy is unusually blue as compared with typical field galaxies. The difference may be due to the fact that the Fruchter et al. (1999a) analysis compared the host colors with a significantly more shallow infrared survey than the NICMOS HDF, essentially masking the trend for faint galaxies to appear more blue.

Figure 5.4 shows a section of the median-binned spectrum of the host galaxy. The Balmer break is clearly detected, with an amplitude $\Delta m \approx 0.55$ mag, which is typical for the Balmer-break-selected population of field galaxies at $z \sim 1$ (K. Adelberger, private communication). For reference we also plot several population-synthesis model spectra (Bruzual & Charlot 1993). The top panel shows model spectra for a galaxy with a uniform star formation rate, which may be a reasonable time-averaged approximation for a normal late-type galaxy. The correspondence is reasonably good and does not depend on the model age. The bottom panel shows models with an instantaneous burst of star formation. In order to match the data, we require fine-tuning of the postburst age to be $\sim 10^8 \times 2^{\pm 1}$ yr. No attempt was made to optimize the fit or to seek best model parameters and the purpose of this comparison is simply illustrative. Clearly, if there was an ongoing or very recent burst of star formation, the spectrum would be much flatter and have a weaker Balmer break.

Section 5.6
Discussion and Conclusion

We have determined the redshift of the presumed host galaxy of GRB 970228 to be $z = 0.695$ based on [O II] 3727 and [O III] 5007 line emission. The implied energy release, $(1.4 \pm 0.3) \times 10^{52}$ erg [20—

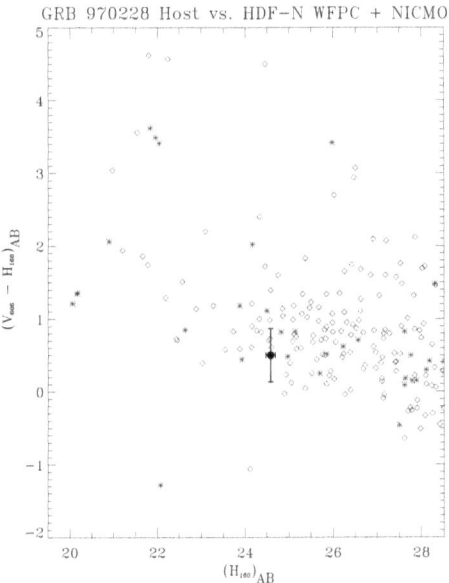

Figure 5.3 Comparison of the color-magnitude of the host galaxy of GRB 970228 with the Hubble Deep Field-North (HDF-N) showing that the host galaxy of GRB 970228 is not especially blue in color. No systematic difference, assuming a Galactic extinction toward GRB 970228 of $A_V = 0.75$ mag, is found between field galaxies at comparable magnitudes and the host (*filled circle with error bars*). NICMOS and WFPC photometry are taken from Thompson et al. (1999) and Williams et al. (1996), respectively. Host galaxy magnitudes and magnitudes of the HDF comparison field objects are aperture-corrected. Diamonds (◇) represent extended objects (with the ratio of semi-major to semi-minor axes less than 0.9) and the asterisks (∗) compact galaxies and stars. The error bars on the HDF-N data have been suppressed. See text for an explanation of the selection criteria.

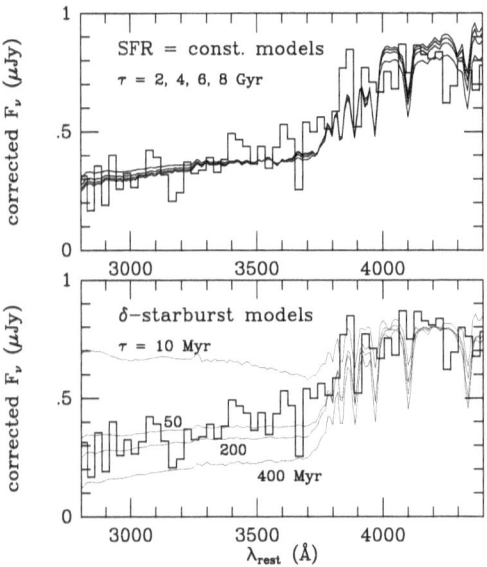

Figure 5.4 Median-binned portion of the host spectrum near the Balmer decrement. *Top panel*: Overlaid are Bruzual & Charlot (1993) galaxy synthesis models assuming a varying time of constant star formation. *Bottom panel*: Overlaid are Bruzual & Charlot (1993) galaxy synthesis models assuming an instantaneous burst of star-formation occurred τ years since observation. Clearly the host continuum could not be dominated by a young population of stars ($\tau = 10$ Myr). See text for a discussion.

2000 keV restframe], is on the smaller end of, but still comparable to, the other bursts with energy determinations (Bloom et al. 2001b). The absence of a detectable break in the afterglow light curve implies that any collimation of emission (i.e., jetting) will not significantly reduce the estimate of total energy release in GRB 970228 (although Frail, Waxman, & Kulkarni 2000a, using late-time radio data, have found that even without an optical break, GRB 970508 may have been collimated).

Most GRB transients appear spatially coincident with faint galaxies, disfavoring the merging neutron star hypothesis (e.g., Paczyński 1998; Bloom et al. 2002). The coincidence of GRB 970508 with its host (Fruchter & Pian 1998; Bloom et al. 1998b) is particularly constraining given the excellent spatial coincidence of the GRB with the center of a dwarf galaxy (see Bloom et al. 1998b). The transient of GRB 970228 lies $3.34 \pm 0.11\ h_{65}^{-1}$ kpc from the center of the galaxy, about 1 kpc in projection outside the half-light radius of the galaxy. From the above analysis we have shown that, like GRB 970508, the host is sub-luminous ($L \approx 0.05 L_*$) and, by assumption, sub-massive relative to L_* galaxies at comparable redshifts. According to Bloom et al. (1999a), about 50% of merging neutron binaries should occur beyond 3.5 kpc in projection of such dwarf galaxies. Thus, by itself, the offset of GRB 970228 from its host does not particularly favor a progenitor model.

In R-band magnitude the host is near the median of GRB hosts observed to date, but in absolute B-band magnitude, the host at the faint end of the distribution. Of the host galaxies detected thus far only GRB 970508 is as comparably faint to the host of GRB 970228. Except in angular extent, the host galaxies of GRB 970228 and GRB 970508 (Bloom et al. 1998b) bear a striking resemblance. Both appear to be sub-luminous ($L \lesssim 0.1 L_*$), compact, and blue. Spectroscopy of both reveal the presence of the [Ne III] 3869 line, indicative of recent very massive star formation. However, such properties are not shared by all of the GRB hosts studied to date. We note too the rather curious trend that the two GRBs themselves appear to have similar properties in that they decay slowly, they are the two least luminous in term of GRB energetics, and do not exhibit evidence of a strong break in the light curve.

The authors thank the generous support of the staff of the W. M. Keck Foundation. This paper has benefited from stimulating conversations with P. van Dokkum and K. Adelberger. We thank M. van Kerkwijk for help during observing and C. Clemens for his use of dark-time observing nights. This paper has been greatly improved by the thoughtful comments of the anonymous referee. JSB gratefully acknowledges the fellowship from the Fannie and John Hertz Foundation. SGD acknowledges partial funding from the Bressler Foundation. This work was supported in part by grants from the NSF and NASA to SRK.

CHAPTER 6

The Observed Offset Distribution of Gamma-Ray Bursts from Their Host Galaxies: A Robust Clue to the Nature of the Progenitors[†]

J. S. BLOOM, S. R. KULKARNI, S. G. DJORGOVSKI

Palomar Observatory 105-24, California Institute of Technology, Pasadena, CA 91125, USA

Abstract

We present a comprehensive study to measure the locations of γ-ray bursts (GRBs) relative to their host galaxies. In total, we find the offsets of 20 long-duration GRBs from their apparent host galaxy centers utilizing ground-based images from Palomar and Keck and space-based images from the Hubble Space Telescope (HST). We discuss in detail how a host galaxy is assigned to an individual GRB and the robustness of the assignment process. The median projected angular (physical) offset is 0.17 arcsec (1.3 kpc). The median offset normalized by the individual host half-light radii is 0.98 suggesting a strong connection of GRB locations with the UV light of their hosts. This provides strong observational evidence for the connection of GRBs to star-formation.

We further compare the observed offset distribution with the predicted burst locations of leading stellar-mass progenitor models. In particular, we compare the observed offset distribution with an exponential disk, a model for the location of collapsars and promptly bursting binaries (e.g., helium star–black hole binaries). The statistical comparison shows good agreement given the simplicity of the model, with the Kolmogorov-Smirnov probability that the observed offsets derive from the model distribution of $P_{\rm KS} = 0.45$. We also compare the observed GRB offsets with the expected offset distribution of delayed merging remnant progenitors (black hole–neutron star and neutron star–neutron star binaries). We find that delayed merging remnant progenitors, insofar as the predicted offset distributions from population synthesis studies are representative, can be ruled out at the 2×10^{-3} level. This is arguably the strongest observational constraint yet against delayed merging remnants as the progenitors of long-duration GRBs. In the course of this study, we have also discovered the putative host galaxies of GRB 990510 and GRB 990308 in archival HST data.

[†] A version of this chapter was first published in *The Astronomical Journal*, 123, p. 1111–1148 (2002). Some of the text in the published version of the introduction was used in chapter 1; the introduction here has thus been abbreviated.

SECTION 6.1
Introduction

Direct associations with other known astrophysical entities is a possible means toward distinguishing between GRB progenitor models. For massive stars, the energy release from the collapse of the core of the star, just as in supernovae, is sufficient to explode the star itself. This may result in a supernova-like explosion at essentially the same time as a GRB. The first apparent evidence of such a supernova associated with a cosmological GRB came with the discovery of a delayed bright red bump in the afterglow light curve of GRB 980326 (Bloom et al. 1999c). The authors interpreted the phenomena as due to the light-curve peak of a supernova at redshift $z \sim 1$. Later, Reichart (1999) and Galama et al. (2000) found similar such red bump in the afterglow of GRB 970228. Merging remnant progenitors models (e.g., BH–NS, NS–NS systems) have difficulty producing these features in a light curve on such long timescales and so the supernova interpretation, if true, would be one of the strongest direct clues that GRBs come from massive star explosions. However, the supernova story is by no means complete. For instance, in only one other GRB (000911) has marginal ($\sim 2\ \sigma$) evidence of a SN signature been found (Lazzati et al. 2001); further, many GRBs do not appear to show any evidence of SNe signatures (e.g., Hjorth et al. 2000). Even the "supernova" observations themselves find plausible alternative explanations (such as dust echoes) that do not strictly require a massive star explosion (Esin & Blandford 2000; Reichart 2001; Waxman & Draine 2000). We note, however, that all other plausible explanations of the observed late-time bumps require high-density environments found most readily in star forming regions.

Chevalier & Li (2000) emphasize that if a GRB comes from a massive star, then the explosion does not take place in a constant density medium, but in a medium enriched by constant mass loss from the stellar winds. One would expect to see signatures of this wind-stratified medium in the afterglow (e.g., bright sub-millimeter emission at early times, increasing "cooling frequency" with time; see Panaitescu & Kumar 2000; Kulkarni et al. 2000). However, afterglow observations have been inconclusive (Kulkarni et al. 2000) with no unambiguous inference of GRB in such a medium to date.

We emphasize that even the connection of GRBs to stellar-mass progenitors has yet to be established. The most compelling arguments we have outlined (e.g., temporal variability) rely on theoretical interpretations of the GRB phenomena. Further, direct observational results (SNe signatures and transient Fe-line emission) are not yet conclusive.

In this paper we examine the observed locations of GRBs with respect to galaxies. We find an unambiguous correlation of GRB locations with the UV light of their hosts, providing strong indirect evidence for the connection of GRBs to stellar-mass progenitors. Beyond this finding, we aim to use the location of GRBs to distinguish between stellar-mass progenitor models. In §6.2 we review the expectations of GRB locations from each progenitor model. Then in §§6.3–6.4 we discuss the instruments, techniques, and expected uncertainties involved in constructing a sample of GRB locations about their host galaxies. In §6.5 we comment on the data reductions specific to each GRB in our sample. The observed distribution is shown and discussed in §6.6 and then statistically compared with the expected offset distribution of leading progenitor models (§6.7). Last, in §6.8 we summarize and discuss our findings.

SECTION 6.2
Location of GRBs as a Clue to Their Origin

How have locations of GRBs within (or outside) galaxies impacted our understanding of the progenitors of GRBs thus far? The first accurate localization (van Paradijs et al. 1997) of a GRB by way of an optical transient afterglow revealed GRB 970228 to be spatially coincident with a faint galaxy (Sahu et al. 1997; Fruchter et al. 1999a; Bloom et al. 2001a). Though the nearby galaxy was faint, van Paradijs et al. (1997) estimated the *a posteriori* probability of a random location on the sky falling so close to a galaxy by chance to be low. As such, the galaxy was identified as the host of GRB 970228. Sahu et al. (1997) further noted that the OT appeared offset from the center of the galaxy thereby calling into question an AGN origin. Soon thereafter Bloom et al. (1998b) found, and then Fruchter & Pian (1998) confirmed, that GRB 970508 was localized very near the center of a dwarf galaxy. Given that underluminous dwarf galaxies have a weaker gravitational potential with which to bind merging remnant binaries, both Paczyński (1998) and Bloom et al. (1998b) noted that the excellent spatial coincidence of the GRB with its putative host found an easier explanation with a massive star progenitor rather than NS–NS binaries.

Once the afterglow fades, one could study in detail its environment (analogous to low-redshift supernovae). Unfortunately, however, the current instrumentation available for GRB observations cannot pinpoint or resolve individual GRB environments on the scale of tens of parsecs unless the GRB occurs at a low redshift ($z \lesssim 0.2$) and the transient afterglow is well-localized. At higher redshifts (as all GRBs localized to-date), only the very largest scales of galactic structure can be resolved (e.g., spiral arms) even by HST. Therefore, the locations of most individual GRBs do not yield much insight into the nature of the progenitors. Instead, the observed *distribution* of GRBs in and around galaxies must be studied as a whole and then compared with the expectations of the various progenitor models. This is the aim of the present study. As we will demonstrate, while not all GRBs are well-localized, the overall distribution of GRB offsets proves to be a robust clue to the nature of the progenitors.

In this paper we present a sample of GRB offset measurements that represents the most comprehensive and uniform set compiled to-date. Every GRB location and host galaxy image has been re-analyzed using the most uniform data available. The compilation is complete with well-studied GRBs until May 2000. Throughout this paper we assume a flat Λ–cosmology (e.g., de Bernardis et al. 2000) with $H_0 = 65$ km s^{-1} Mpc^{-1}, $\Omega_M = 0.3$, and $\Lambda_0 = 0.7$.

SECTION 6.3
The Data: Selection and Reduction

The primary goal of this paper is to measure the offsets of GRBs from their hosts where the necessary data are available. Ideally this could be accomplished using a dataset of early-time afterglow and late-time host imaging observed using the same instrument under similar observing conditions. The natural instrument of choice is HST given its exquisite angular resolution and astrometric stability. Though while most hosts have been observed with HST at late-times, there are only a handful of early-time HST detections of GRB afterglow. On the other hand, early ground-based images of GRB afterglows are copious but late-time seeing-limited images of the hosts give an incomplete view of the host as compared to an HST image of the same field. Moreover, ground-based imaging is inherently heterogeneous, taken with different instruments, at different signal-to-noise levels, and through a variety observing of conditions; this generally leads to poorer astrometric accuracy. Bearing these imperfections in mind we have compiled a dataset of images that we believe are best suited to find offsets of GRBs from their hosts.

6.3. THE DATA: SELECTION AND REDUCTION

Table 6.1. GRB Host and Astrometry Observing Log

Name (1)	Teles./Instr./Filter (2)	Date (3)	α (J2000) (4)	δ (J2000) (4)	Exp. (5)	Δt (6)	Level (7)	Refs. (8)
GRB 970228...	HST/STIS/049001040	4.75 Sep 1997	05 01 46.7	+11 46 54	4600	189	self-HST	1, 2
GRB 970508...	HST/STIS/041C01DIM	2.64 Jun 1997	06 53 49.5	+79 16 20	5000	25	HST→HST	3, 4
	HST/STIS/04IB01I9Q	6.01 Aug 1997			11568	89		5
GRB 970828...	Keck/LRIS/R-band	19.4 Jul 1998	18 08 34.2	+59 18 52	600	325	RADIO→OPT	6, 7
GRB 971214...	Keck/LRIS/J-band	16.52 Dec 1997	11 56 26.0	+65 12 00	1080	1.5	GB→HST	8
	HST/STIS/04T301040	13.27 Apr 1998			11862	119		9
GRB 980326...	Keck/LRIS/R-band	28.25 Mar 1998	08 36 34.3	−18 51 24	240	1.4	GB→GB	10,11
	Keck/LRIS/R-band	18.50 Dec 1998			2400	267		11
	HST/STIS/059251ZWQ	31.80 Dec 2000			7080	1010		12
GRB 980329...	Keck/NIRC/K-band	2.31 Apr 1998	07 02 38.0	+38 50 44	2520	4.15	GB→GB	13, a
	Keck/ESI/R-band	1.41 Jan 2001			6600	1009		a
	HST/STIS/065K22YXQ	27.03 Aug 2000			8012	884		14
SN 1998bw....	NTT/EMMI/I-band	4.41 May 1998	19 35 03.3	−52 50 45	120	8.5	GB→HST	15
(GRB 980425?)	HST/STIS/065K30B1Q	11.98 Jun 2000			1185	778		16
GRB 980519...	P200/COSMIC/R-band	20.48 May 1998	23 22 21.5	+77 15 43	480	1.0	GB→GB→HST	17, 18
	Keck/LRIS/R-band	24.50 Aug 1998			2100	97		19
	HST/STIS/065K41IEQ	7.24 Jun 2000			8924	750		20
GRB 980613...	Keck/LRIS/R-band	16.29 Jun 1998	10 17 57.6	+71 27 26	600	3.1	GB→GB→HST	21, a
	Keck/LRIS/R-band	29.62 Nov 1998			900	169		a
	HST/STIS/065K51ZZQ	20.31 Aug 2000			5851	799		22
GRB 980703...	Keck/LRIS/R-band	6.61 Jul 1998	23 59 06.7	+08 35 07	600	3.4	GB→HST	23
	HST/STIS/065K61XTQ	18.81 Jun 2000			5118	717		24
GRB 981226...	Keck/LRIS/R-band	21.57 Jun 1999	23 29 37.2	−23 55 54	3360	177	RADIO→OPT(GB)→HST	25
	HST/STIS/065K71AXQ	3.56 Jul 2000			8265	555		26
GRB 990123...	HST/STIS/059601060	9.12 Feb 1999	15 25 30.3	+44 45 59	7200	16.7	self-HST	27
GRB 990308...	Keck/LRIS/R-band	19.26 Jun 1999	12 23 11.4	+06 44 05	1000	103	GB→GB→HST	28
	HST/STIS/065K91E6Q	19.67 Jun 2000			7782	470		29
GRB 990506...	Keck/LRIS/R-band	11.25 Jun 1999	11 54 50.1	−26 40 35	1560	36	RADIO→OPT(GB)→HST	30
	HST/STIS/065KA1UVQ	24.55 Jun 2000			7856	415		31
GRB 990510...	HST/STIS/059273LCQ	17.95 Jun 1999	13 38 07.7	−80 29 49	7440	39	HST→HST	32, 33
	HST/STIS/059276C7Q	29.45 Apr 2000			5840	355		34
GRB 990705...	NTT/SOFI/H-band	5.90 Jul 1999	05 09 54.5	−72 07 53	1200	0.23	GB→GB→HST	35
	VLT/FORS1/V-band	10.40 Jul 1999			1800	4.7		35
	HST/STIS/065KB1G2Q	26.06 Jul 2000			8792	386		36
GRB 990712...	HST/STIS/059262VEQ	29.50 Aug 1999	22 31 53.1	−73 24 28	8160	48	HST→HST	37, 38
	HST/STIS/059274BNQ	24.21 Apr 2000			3720	287		39

A listing of the dataset compilation is given in table 6.2. We include every GRB (up to and including GRB 000418) with an accurate radio or optical location and a deep late-time optical image. There is a hierarchy of preference of imaging conditions and instruments which yield the most accurate offsets; we describe the specifics and expected accuracies of the astrometric technique in §6.4.

6.3.1 Dataset selection based on expected astrometric accuracy

We group the datasets into five different levels ordered by decreasing astrometric accuracy. Levels 1–4 each utilize differential astrometry and level 5 utilizes absolute astrometry relative to the International Coordinate Reference System (ICRS). Specifics of the individual offset measurements are given in §6.4. The ideal dataset for offset determination is a single HST image where both the transient and the host are well-localized (hereafter "self-HST"); so far, only GRB 970228, GRB 990123 and possibly GRB 991216 fall in this category. The next most accurate offset is obtained where both the early- and late-time images are from HST taken at comparable depth with the same filter (hereafter "HST→HST"). In addition to the centering errors of the OT and host, such a set inherits the uncertainty in registering the two epochs (e.g., GRB 970508). Next, an early deep image from ground-based (GB) Keck, Palomar 200-inch (P200), or the Very Large Telescope (VLT) in which the OT dominates is paired with a late-time image from HST (e.g., GRB 971214, GRB 980703, GRB 991216, GRB 000418; "GB→HST"). Though in the majority of these cases most of the objects detected in the HST image are also detected in the Keck image (affording great redundancy in the astrometric mapping solution), object centering of ground-based data is hampered by atmospheric seeing. The next most accurate localizations use ground-based to ground-based imaging to compute offsets ("GB→GB"). Last, radio localizations compared with optical imaging ("RADIO→OPT") provide the least accurate offset determinations. This is due primarily to the current difficulty of mapping an optical image onto an absolute coordinate system (see §6.4.5).

6.3.2 Imaging reductions

Reductions of HST Imaging

Most of the HST images of GRB afterglows and hosts were acquired using the *Space Telescope Imaging Spectrograph* (STIS; Kimble et al. 1998). STIS imaging under-samples the angular diffraction limit of the telescope and therefore individual HST images essentially do not contain the full astrometric information possible. To produce a final image that is closer to the diffraction limit, inter-pixel dithering between multiple exposures is often employed. The image reconstruction technique, which also facilitates removal of cosmic-rays and corrects for the known optical field distortion, is called "drizzling" and is described in detail in Fruchter & Hook (1997). We use this technique, as implemented using the IRAF[1] package DITHER and DITHERII, to produce our final HST images.

We retrieved and reduced every public STIS dataset of GRB imaging from the HST archive[2] and processed the so-called "on-the-fly calibration" images to produce a final drizzled image. These images are reduced through the standard HST pipeline for bias subtraction, flat-fielding, and illumination corrections using the best calibration data available at the time of archive retrieval. The archive name of the last image and the start time of each HST epoch are given in columns 2 and 3 of table 6.1.

Some HST GRB imaging has been taken using the STIS/Longpass filter (F28x50LP) which, based on its red effective wavelength (central wavelength $\lambda_c \approx 7100$Å), would make for a good comparison with ground-based R-band imaging. However, the Longpass filter truncates the full STIS field of view to about 40% and therefore systematically contains fewer objects to tie astrometrically to ground-based images. Therefore, all of the HST imaging reported herein were taken in (unfiltered) STIS/Clear (CCD50) mode. Unlike the Longpass filter, the spectral response of the Clear mode is rather broad

---
[1] IRAF is distributed by the National Optical Astronomy Observatories, which are operated by the Association of Universities for Research in Astronomy, Inc., under cooperative agreement with the National Science Foundation.
[2] http://archive.stsci.edu

Table 6.1—Continued

Name (1)	Teles./Instr./Filter (2)	Date (3)	α (J2000) (4)	δ (J2000) (4)	Exp. (5)	Δt (6)	Level (7)	Refs. (8)
GRB 991208...	Keck/NIRSPEC/K-band	16.68 Dec 1999	16 33 53.5	+46 27 21	1560	8.5	GB→GB→HST	40, 41
	Keck/ESI/R-band	4.54 Apr 2000			1260	118		a
	HST/STIS/O5926 6ODQ	3.58 Aug 2000			5120	239		42
GRB 991216...	Keck/ESI/R-band	29.41 Dec 1999	05 09 31.2	+11 17 07	600	13	GB→GB→HST	43, a
	Keck/ESI/R-band	4.23 Apr 2000			2600	110		a
	HST/STIS/O59272GIQ	17.71 Apr 2000			9440	123		44
GRB 000301C...	HST/STIS/O59277P9Q	6.22 Mar 2000	16 20 18.6	+29 26 36	1440	4.8	HST→HST	45, 46
	HST/STIS/O59265XYQ	25.86 Feb 2001			7361	361		47
GRB 000418...	Keck/ESI/R-band	28.41 Apr 2000	12 25 19.3	+20 06 11	300	10	GB→HST	48, 49
	HST/STIS/O59264Y6Q	4.23 Jun 2000			2500	47		50

Note. — (2) Telescopes: HST = *Hubble Space Telescope* Keck = W. M. Keck 10 m Telescope II, Mauna Kea, Hawaii, P200 = Hale 200-inch Telescope at Palomar Observatory, Palomar Mountain, California, NTT = European Space Agency 3.5 m New Technology Telescope, Chile, VLT = Very Large Telescope UT-1 ("Antu"); Instruments: STIS (Kimble et al. 1998), ESI (Epps & Miller 1998), LRIS (Oke et al. 1995), COSMIC (Kells et al. 1998), NIRSPEC (McLean et al. 1998), SOFI (Finger et al. 1998), FORS1 (Nicklas et al. 1997); Filter: all ground-based observations are listed in standard bandpass filters while the HST/STIS images (used for astrometry) are all in Clear Mode. The last dataset of the HST visit is listed. (3) Observation dates in Universal Time (UT) corresponding to the start time of the last observation in the dataset. (4) Position (α: hours, minutes, seconds and δ: degrees, arcminutes, and arcseconds) of the GRB. (5) Total exposure time in seconds. (6) Time in days since the trigger time of the GRB. (7) The comment denotes the astrometric level as in §6.4. (8) Reference to the first presentation of the given dataset. If two references appear on a given line then the first is a reference to the position of the GRB.

References. — a. This paper; 1. van Paradijs et al. (1997); 2. Fruchter et al. (1999a); 3. Frail et al. (1997); 4. Pian et al. (1998); 5. Fruchter & Pian (1998); 6. Groot et al. (1998b); 7. Djorgovski et al. (2001); 8. Kulkarni et al. (1998b); 9. Odewahn et al. (1998); 10. Groot et al. (1998a); 11. Bloom et al. (1998c); 12. Fruchter et al. (2001a); 13. Larkin et al. (1998); 14. Holland et al. (2000g); 15. Galama et al. (1998b); 16. Holland et al. (2000b); 17. Jaunsen et al. (1998); 18. Bloom et al. (1998a); 19. Bloom et al. (1998d); 20. Holland et al. (2000c); 21. Hjorth et al. (1998); 22. Holland et al. (2000f); 23. Bloom et al. (1998b); 24. Bloom et al. (2000a); 25. Frail et al. (1999); 26. Hjorth et al. (2000); 27. Bloom et al. (1999b); 28. Schaefer et al. (1999); 29. Holland et al. (2000e); 30. Taylor et al. (2000); 31. Holland et al. (2000d); 32. Vreeswijk et al. (1999); 33. Fruchter et al. (1999); 34. Bloom (2000); 35. Masetti et al. (2000); 36. Holland et al. (2000a); 37. Sahu et al. (2000); 38. Fruchter et al. (2000b); 39. Fruchter et al. (2000c); 40. Frail (1999); 41. Bloom et al. (1999); 42. Fruchter et al. (2000d); 43. Uglesich et al. (1999); 44. Vreeswijk et al. (2000); 45. Fynbo et al. (2000); 46. Fruchter et al. (2000e); 47. Fruchter et al. (2001b); 48. Mirabal et al. (2000); 49. Bloom et al. (2000b); 50. Metzger et al. (2000)

Table 6.2. Measured Angular Offsets and Physical Projections

Name	X_0 East "	Y_0 North "	R_0 "	R_0/σ_{R_0}	z	D_θ kpc/"	X_0 (proj) kpc	Y_0 (proj) kpc	R_0 (proj) kpc
GRB 970228	−0.033±0.034	−0.424±0.034	0.426±0.034	12.59	0.695	7.673	−0.251±0.259	−3.256±0.259	3.266±0.259
GRB 970508	0.011±0.011	0.001±0.012	0.011±0.011	1.003	0.835	8.201	0.090±0.090	0.008±0.098	0.091±0.090
GRB 970828	0.440±0.516	0.177±0.447	0.474±0.507	0.936	0.958	8.534	3.755±4.403	1.510±3.815	4.047±4.326
GRB 971214	0.120±0.070	−0.070±0.070	0.139±0.070	1.985	3.418	7.952	0.954±0.557	−0.557±0.557	1.105±0.557
GRB 980326	0.125±0.068	−0.037±0.062	0.130±0.068	1.930	∼1	...	...	...	...
GRB 980329	−0.037±0.049	−0.003±0.061	0.037±0.049	0.756	≲3.5	...	...	...	...
GRB 980425	−10.55±0.052	−6.798±0.052	12.55±0.052	241.4	0.008	0.186	−1.964±0.010	−1.265±0.010	2.337±0.010
GRB 980519	−0.050±0.130	1.100±0.100	1.101±0.100	11.00	...	...	...	...	...
GRB 980613	0.039±0.052	0.080±0.080	0.089±0.076	1.174	1.096	8.796	0.344±0.454	0.703±0.707	0.782±0.666
GRB 980703[a]	−0.054±0.055	0.098±0.065	0.112±0.063	1.788	0.966	8.553	−0.460±0.469	0.842±0.555	0.959±0.536
GRB 981226	0.616±0.361	0.426±0.246	0.749±0.328	2.282	...	...	...	...	...
GRB 990123	−0.192±0.003	−0.641±0.003	0.669±0.003	223.0	1.600	9.124	−1.752±0.027	−5.849±0.027	6.105±0.027
GRB 990308	−0.328±0.357	−0.989±0.357	1.042±0.357	2.919	...	...	...	...	...
GRB 990506	−0.246±0.432	0.166±0.513	0.297±0.459	0.647	1.310	9.030	−2.221±3.901	1.499±4.632	2.680±4.144
GRB 990510	−0.064±0.009	0.015±0.012	0.066±0.009	7.160	1.619	9.124	−0.584±0.082	0.137±0.109	0.600±0.084
GRB 990705	−0.865±0.046	0.109±0.049	0.872±0.046	18.86	0.840	8.217	−7.109±0.380	0.896±0.399	7.165±0.380
GRB 990712	−0.035±0.080	0.035±0.080	0.049±0.080	0.619	0.434	6.072	−0.213±0.486	0.213±0.486	0.301±0.486
GRB 991208	0.071±0.102	0.183±0.096	0.196±0.097	2.016	0.706	7.720	0.548±0.789	1.410±0.744	1.513±0.750
GRB 991216	0.211±0.029	0.290±0.034	0.359±0.032	11.08	1.020	8.664	1.828±0.251	2.513±0.295	3.107±0.280
GRB 000301C	−0.025±0.014	−0.065±0.005	0.069±0.007	9.821	2.030	9.000	−0.222±0.130	−0.581±0.046	0.622±0.063
GRB 000418	−0.019±0.066	0.012±0.058	0.023±0.064	0.358	1.118	8.829	−0.170±0.584	0.109±0.514	0.202±0.564

[a]Using radio detections of the host and afterglow, Berger et al. (2001a) find a more accurate offset of $X_0 = -0.032 \pm 0.015$ and $Y_0 = 0.025 \pm 0.015$ ($R_0 = 0.040 \pm 0.015$), consistent with our optical results; see §6.5.10.

Note. — The observed offsets (X_0, Y_0) and associated Gaussian uncertainties include all statistical errors from the astrometric mapping and OT+host centroid measurements. The observed offset, $R_0 = \sqrt{X_0^2 + Y_0^2}$, (constructed analogously to equation 6.8) are given in col. 4. Note that $R_0 - \sigma_{R_0} \leq R \leq R_0 + \sigma_{R_0}$ is not necessarily the 68% percent confidence region of the true offset since the probability distribution is not Gaussian (see equation 6.7). The term R_0/σ_{R_0} in col. 5 indicates how well the offset from the host center is determined. In general, we consider the GRB to be significantly displaced from the host center if $R_0/\sigma_{R_0} \gtrsim 5$. In col. 6, z is the measured redshift of the host galaxy and/or absorption redshift of the GRB complied from the literature (see Kulkarni et al. 2000). In col. 7, D_θ is the conversion of angular displacement in arcseconds to projected physical distance.

(2000–10000 Å). We use the known optical distortion coefficients appropriate to the wavelength of peak sensitivity $\lambda \approx 5850$ Å of this observing mode to produce final images which are essentially linear in angular displacement versus instrumental pixel location.

The original plate scale of most STIS imaging is $0''.05077 \pm 0.00007$ pixel^{-1} (Malumuth & Bowers 1997), though there is a possibility that thermal expansion of the instrument could change this scale by a small amount (see Appendix 6.A). The pixel scale of all our final reduced HST images is half the original scale, i.e., $0''.02539$ pixel^{-1}.

Reductions of Ground-based Imaging

Ground-based images are all reduced using the standard practices for bias subtraction, flat-fielding, and in the case of I-band imaging, fringe correction. In constructing a final image we compute the instrumental shift of dithered exposures relative to a fiducial exposure and co-add the exposures after applying the appropriate shift to align each image. All images are visually inspected for cosmic-ray contamination of the transient, host or astrometric tie stars. Pixels contaminated by cosmic rays are masked and not used in the production of the final image.

SECTION 6.4
Astrometric Reductions and Issues Related to Dataset Levels

Here we provide a description of the astrometric reduction techniques for both our ground-based and the HST images, and issues related to the five levels of astrometry summarized in §6.3.1. A discussion of the imaging reductions and astrometry for the individual cases is given in section §6.5.

6.4.1 Level 1: self-HST (differential)

An ideal image is one where the optical transient and the host galaxy are visible in the same imaging epoch with HST. This typically implies that the host galaxy is large enough in extent to be well-resolved despite the brilliance of the nearby OT. Of course, a later image of the host is always helpful to confirm that the putative afterglow point source does indeed fade. In this case (as with GRB 970228 and GRB 990123) the accuracy of offset determination is limited mostly by the centroiding errors of the host "center" and optical afterglow. Uncertainties in the optical distortion corrections and the resulting plate scale are typically sub-milliarcsecond in size (see Appendix 6.A).

In principle we expect centering techniques to result in centroiding errors (σ_c) on a point source with a signal-to-noise, SN, of $\sigma_c \approx \phi/SN$ (see Stone 1989), where ϕ is the instrumental full-width half-maximum (FWHM) seeing of the final image. Since ϕ is typically ~ 75 milliarcsecond (mas) we expect $\sim$milliarcsecond offset accuracies with self-HST images.

6.4.2 Level 2: HST→HST (differential)

Here, two separate HST epochs are used for the offset determination. The first epoch is taken when the afterglow dominates the light and the second when the host dominates. In addition to the centroiding errors, the astrometric accuracy of this level is limited by uncertainty in the registration between the two images.

In general when two images are involved (here and all subsequent levels), we register the two images such that an instrumental position in one image is mapped to the instrumental (or absolute position) in the other image. The registration process is as follows. We determine the noise characteristics of both of the initial and final images empirically, using an iterative sigma-clipping algorithm. This noise along with the gain and effective read noise of the CCD are used as input to the IRAF/CENTER algorithm. In addition we measure the radial profile of several apparent compact sources in the image and use the derived seeing FWHM (ϕ) as further input to the optimal filtering algorithm technique for centering

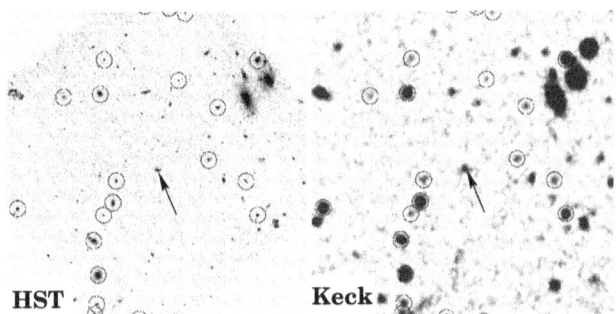

Figure 6.1 Example Keck R-band and HST/STIS Clear images of the field of GRB 981226. Twenty of the 25 astrometric tie objects are circled in both images. As with other Keck images used for astrometry in the present study, most of the faint object detected in the HST images are also detected (albeit with poorer resolution). The optical transient in the Keck image and the host galaxy in the HST image are in center. The field is approximately $50'' \times 50''$ with North up and East to the left.

(OFILTER; see Davis 1987). For faint stars we use the more stable GAUSSIAN algorithm. Both techniques assume a Gaussian form of the point-spread function which, while not strictly matched to the outer wings of the Keck or HST point-spread functions (PSFs), appears to reasonably approximate the PSF out to the FWHM of the images.

When computing the differential astrometric mappings between two images (such as HST and Keck or Keck and the USNO-A2.0 catalog), we use a list of objects common from both epochs, "tie objects," and compute the astrometric mapping using the routine IRAF/GEOMAP. The polynomial order of the differential fitting we use depends on the number of tie objects. A minimum of three tie objects are required to find the relative rotation, shift and scale of two images, which leaves only one degree of freedom. The situation is never this bad; in fact, when comparing HST images and an earlier HST image (or deep Keck image), we typically find 20–30 reasonable tie objects, and therefore we can solve for higher-order distortion terms. Figure 6.1 shows an example Keck and HST field of GRB 981226 and the tie objects we use for the mapping. We always reject tie objects that deviate by more than 3σ from the initial mapping. A full third-order two-dimensional polynomial with cross-terms requires 18 parameters which leaves, typically, $N \approx 30$ degrees of freedom. Assuming such a mapping adequately characterizes the relative distortion, and it is reasonable to expect that mapping errors will have an r.m.s. error $\sigma \approx 30^{-1/2}\phi/\langle SN\rangle$, where $\langle SN\rangle$ is the average signal-to-noise of the tie objects. For example, in drizzled HST images $\phi \approx 75$ mas and $\langle SN\rangle \approx 20$ so that we can expect differential mapping uncertainties at the 1 mas level for HST→HST mapping. Cross-correlation techniques, such as IRAF/CROSSDRIZZLE, can in principle result in even better mapping uncertainties, but, in light of recent work by Anderson & King (1999), we are not confident that the HST CCD distortions can be reliably removed at the sub-mas level.

6.4.3 Level 3: GD→IIST (differential)

This level of astrometry accounts for the majority of our dataset. In addition to inheriting the uncertainties of centroiding errors and astrometric mapping errors described above, we must also consider the effects of differential chromatic refraction (DCR) and optical image distortion in ground based images. In Appendix 6.A we demonstrate that these effects should not dominate the offset uncertainties. Following the argument above (§6.4.2) the astrometric mapping uncertainties scale linearly with the seeing of ground-based images which is typically a factor of 10–20 larger than the effective seeing of the HST images.

An independent test of the accuracy of the transference of differential astrometry from ground-based images to space-based imaging is illustrated by the case of GRB 990123. In Bloom et al. (1999b) we registered a Palomar 60-inch (P60) image to a Keck image and thence to an HST image. The overall statistical uncertainty introduced by this process (see Appendix 6.B for a derivation) is $\sigma_r = 107$ mas (note that in the original paper we mistakenly overstated this error as 180 mas uncertainty). The position we inferred was 90 mas from a bright point source in the HST image. This source was later seen to fade in subsequent HST imaging and so our identification of the source as the afterglow from P60→Keck→HST astrometry was vindicated. Since the P60→Keck differential mapping accounted for approximately half of the error (due to optical field distortion and unfavorable seeing in P60), we consider ~ 100 mas uncertainty in Keck→HST mapping as a reasonable upper limit to the expected uncertainty from other cases. In practice, we achieved r.m.s. accuracies of 40–70 mas (see table 6.2).

6.4.4 Level 4: GB→GB (differential)

This level contains the same error contributions as in GB→HST level, but in general, the uncertainties are larger since the centroiding uncertainties are large in both epochs. The offsets are computed in term of pixels in late-time image. Just as with the previous levels with HST, we assume an average plate scale to convert the offset to units of arcseconds. For the Low Resolution Imaging Spectrometer (LRIS; Oke et al. 1995) and the Echellete Spectrograph Imager (ESI; Epps & Miller 1998) imaging, we assume a plate scale of $0''.212$ pixel^{-1} and $0''.153$ pixel^{-1}, respectively. We have found that these plate scales are stable over time to better than a few percent; consequently, the errors introduced by any deviations from these assumed plate scales are negligible.

6.4.5 Level 5: RADIO→OPT (absolute)

Unfortunately, the accuracy of absolute offset determination is (currently) hampered by systematics in astrometrically mapping deep optical/infrared imaging to the ICRS. Only bright stars ($V \lesssim 9$ mag) have absolute localizations measured on the milliarcsecond level thanks to astrometric satellite missions such as Hipparcos. The density of Hipparcos stars is a few per square degree so the probability of having at least two such stars on a typical CCD frame is low. Instead, optical astrometric mapping to the ICRS currently utilizes ICRS positions of stars from the USNO-A2.0 Catalog, determined from scanned photographic plates (Monet 1998). Even if all statistical errors of positions are suppressed, an astrometric plate solution can do no better than inherit a systematic 1-σ uncertainty of 250 mas in the absolute position of any object on the sky ($\sigma_\alpha = 0''.18$ and $\sigma_\delta = 0''.17$; Deutsch 1999). By contrast, very-long baseline array (VLBA) positions of GRB radio afterglow have achieved sub-milliarcsecond absolute positional uncertainties relative to the ICRS (Taylor et al. 1999). So, until optical systematics are beaten down and/or sensitivities at radio wavelengths are greatly improved (so as to directly detect the host galaxy at radio wavelengths), the absolute offset astrometry can achieve 1-σ accuracies no better than ~ 300 mas (≈ 2.5 kpc at $z = 1$). In fact, one GRB host (GRB 980703) has been detected at radio wavelengths (Berger et al. 2001a) with the subsequent offset measurement accuracy improving by a factor of ~ 3 over the optical measurement determined herein (§6.5.10).

There are three GRBs in our sample where absolute astrometry (level 5) is employed. In computing the location of the optical transient relative to the ICRS we typically use 20–40 USNO A2.0 astrometric tie stars in common with Keck or Palomar images. We then use IRAF/CCMAP to compute the mapping of instrumental position (x, y) to the world coordinate system (α, δ).

SECTION 6.5
Individual Offsets and Hosts

Below we highlight the specific reductions for each offset, the results of which are summarized in table 6.2. In total there are 21 bursts until May 2000 that have been reliably localized at the arcsecond level and 1 burst with an uncertain association with the nearby SN 1998bw (GRB 980425). In our analysis we do not include GRB 000210 (Stornelli et al. 2000) due to lack of late-time imaging data. Thereby, the present study includes 20 "cosmological" GRBs plus the nearby SN 1998bw/GRB 980425. Offset measurements should be possible for the recent bursts GRB 000630 (Hurley et al. 2000a), GRB 000911 (Hurley et al. 2000b), GRB 000926 (Hurley et al. 2000c) and GRB 010222 (Piro et al. 2001b).

To look for the hosts, we generally image each GRB field roughly a few months to a year after the burst with Keck. Typically, these observations reach a limiting magnitude of $R \approx 24$–26 mag depending on the specifics of the observing conditions. If an object is detected within ~ 1 arcsec from the afterglow position and has a brightness significantly above the extrapolated afterglow flux at the time of observation, this source is deemed the host (most GRB hosts are readily identified in such imaging). If no object is detected, we endeavor to obtain significantly deeper images of the field. Typically these faint host searches require 1–3 hours of Keck (or VLT) imaging to reach limiting magnitudes at the $R \approx 27$ mag level. If no object is detected at the location of the afterglow, HST imaging is required and the host search is extended to limiting magnitudes of $R \approx 28$–29 mag. Only 3 hosts in our sample (GRB 990510, GRB 000301C, and GRB 980326) were first found using HST after an exhaustive search from the ground.

Note that the assignment of a certain observed galaxy as the host of a GRB is, to some extent, a subjective process. We address the question of whether our assignments are "correct" in §6.6.1 where we demonstrate on statistical grounds that at most only a few assignments in the sample of 20 could be spurious. In §6.6.1 we also discuss how absorption/emission redshifts help strengthen the physical connection of GRBs to their assigned hosts.

Irrespective of whether individual assignments of hosts are correct, we uniformly assign the nearest (in angular distance) detected galaxy as the host. In practice this means that the nearest object (i.e., galaxy) brighter than $R \simeq 25$–26 mag detected in Keck imaging is assigned as the host. In almost all cases, there is a detected galaxy within ~ 1 arcsecond of the transient position. For the few cases where there is no object within ~ 1 arcsecond, deeper HST imaging *always* reveals a faint galaxy within ~ 1 arcsecond. In most cases, the estimated probability that we have assigned the "wrong" host is small (see §6.6.1). After assigning the host, the center of host is then determined, except in a few cases, as the centroid near the brightest component of the host system. In a few cases where there is evidence for significant low-surface brightness emission (e.g., 980519) or the host center is ambiguous, we assign the approximate geometric center as the host center.

A summary of our offset results is presented in table 6.2. Since all our final images of the host galaxies are rotated to the cardinal orientation before starting the astrometric mapping process, these uncertainties are also directly proportional to the uncertainties in α and δ. It is important to note, however, that the projected radial offset is a positive-definite number and the probability distribution is not Gaussian. Thereby, the associated error (σ_r) in offset measurements does not necessarily yield a 68% confidence region for the offset (see Appendix 6.B) but is, clearly, indicative of the precision of the offset measurement.

Once the offsets are determined from the final images, we then measure the half-light radii of the host galaxies. For extended hosts, the value of the half-light radius may be obtained directly from aperture curve-of-growth analysis. However, for compact hosts, the instrumental resolution systematically spreads the host flux over a larger area and biases the measurement of the half-light radius to larger values. We attempt to correct for this effect (for all hosts, not just compact hosts) by deconvolving the images with IRAF/SCLEAN using an average STIS/Clear PSF derived from 10 stars in the final HST image of the GRB 990705 field (which were obtained through low Galactic latitude). We then fit curve-of-growth photometry about the host centers and determine the radius at which half the detected light was within such radius. These values, along with associated errors are presented in table 6.3. We tested that the PSFs derived at differing roll angles and epochs had little impact upon the determined value of the half-light radius.

6.5.1 GRB 970228

The morphology and offset derivation have been discussed extensively in Bloom et al. (2001a) and we briefly summarize the results. In the HST/STIS image (figure 6.2), the host appears to be essentially a face-on late-type blue dwarf galaxy. At the center is an apparent nucleus manifested as a 6-σ peak north of the transient. There is also an indication of arm-like structure extending toward the transient.

This image represents the ideal for astrometric purposes (level 1): both the transient and the host "center" are well-localizable in the same high-resolution image. The transient appears outside the half-light radius of the galaxy.

6.5.2 GRB 970508

The host is a compact, elongated and blue galaxy (Bloom et al. 1998b) and is likely undergoing a starburst phase. The optical transient was well-detected in the early time HST image (Pian et al. 1998) and the host was well-detected (figure 6.2) in the late-time image (Fruchter & Pian 1998). We masked out a $2'' \times 2''$ region around the OT/host and cross-correlated the two final images using the IRAF/CROSSDRIZZLE routine. We used the IRAF/SHIFTFIND routine on the correlation image to find the systematic shift between the two epochs. The resulting uncertainty in the shift was quite small, $\sigma = 0.013, 0.011$ pix (x, y direction). We also found 37 compact objects in common to both images and performed an astrometric mapping in the usual manner. We find $\sigma = 0.344, 0.354$ pix in the (x,y directions). We centered the OT and the host in the normal manner using the IRAF/ELLIPSE task.

The resulting offset is given in table 6.2 where we use the more conservative astrometric mapping uncertainties from using the tie objects, rather than the CROSSDRIZZLE routine. As first noted in Bloom et al. (1998b) (Keck imaging) and then in Fruchter & Pian (1998) (HST imaging), the OT was remarkably close to the apparent center of the host galaxy. The P200→Keck astrometry from Bloom et al. (1998b) produced an r.m.s. astrometric uncertainty of 121 mas, compared to an r.m.s. uncertainty of 11 mas from HST→HST astrometry. The largest source of uncertainty from the HST→HST is the centroid position of the host galaxy.

6.5.3 GRB 970828

The host is identified as the middle galaxy in an apparent three-component system. We discuss the host properties and the astrometry (RADIO→OPT) in more detail in Djorgovski et al. (2001). The total uncertainty in the radio to Keck tie is 506 mas (α) and 376 mas (δ).

6.5.4 GRB 971214

By all accounts, the host appears to be a typical L_* galaxy at redshift $z = 3.42$. The Keck→HST astrometry is discussed in detail in Odewahn et al. (1998). The offset uncertainty found was $\sigma_r = 70$ mas. The GRB appears located to the east of the host galaxy center, but consistent with the east-west extension of the host (see figure 6.2).

6.5.5 GRB 980326

No spectroscopic redshift for this burst was found. However, based on the light-curve and the SN hypothesis, the presumed redshift is $z \sim 1$ (see Bloom et al. 1999c). Bloom et al. (1999c) reported that no galaxy was found at the position of the optical transient down to a 3-σ limiting magnitude

Table 6.3. Host Detection Probabilities and Host Normalized Offsets

Name	$R_{c,host}$ mag	A_{R_c} mag	P_{chance}	R_{half} (obs) "	R_{half} (calc) "	r_e kpc	r_0
GRB 970228	24.60±0.20	0.630	0.00935	0.345±0.030	0.316±0.095	1.6	1.233±0.146
GRB 970508	24.99±0.17	0.130	0.00090	0.089±0.026	0.300±0.090	0.4	0.124±0.129
GRB 970828	25.10±0.30	0.100	0.07037	...	0.296±0.089	1.5	1.603±1.780
GRB 971214	25.65±0.30	0.040	0.01119	0.226±0.031	0.273±0.082	1.1	0.615±0.321
GRB 980326	28.70±0.30	0.210	0.01878	0.043±0.028	0.116±0.035	0.2	3.023±2.532
GRB 980329	27.80±0.30	0.190	0.05493	0.245±0.033	0.168±0.050	1.3	0.152±0.202
GRB 980425	14.11±0.05	0.170	0.00988	18.700±0.025	...	2.1	0.671±0.003
GRB 980519	25.50±0.30	0.690	0.05213	0.434±0.041	0.279±0.084	2.2	2.540±0.332
GRB 980613	23.58±0.10	0.230	0.00189	0.227±0.031	0.352±0.106	1.2	0.392±0.338
GRB 980703	22.30±0.08	0.150	0.00045	0.169±0.026	0.392±0.117	0.9	0.663±0.385
GRB 981226	24.30±0.01	0.060	0.01766	0.336±0.030	0.327±0.098	1.7	2.227±0.996
GRB 990123	23.90±0.10	0.040	0.01418	0.400±0.028	0.341±0.102	2.2	1.673±0.117
GRB 990308	28.00±0.50	0.070	0.31659	0.213±0.028	0.156±0.047	1.1	4.887±1.776
GRB 990506	24.80±0.30	0.180	0.04365	0.090±0.027	0.308±0.092	0.5	3.297±5.196
GRB 990510	27.10±0.30	0.530	0.01218	0.167±0.041	0.205±0.061	0.9	0.393±0.111
GRB 990705	22.00±0.10	0.334[a]	0.01460	1.151±0.030	0.400±0.120	5.7	0.758±0.045
GRB 990712	21.90±0.15	0.080	0.00088	0.282±0.026	0.403±0.121	1.0	0.175±0.284
GRB 991208	24.20±0.20	0.040	0.00140	0.048±0.026	0.330±0.099	0.2	4.083±2.994
GRB 991216	25.30±0.20	1.640	0.00860	0.400±0.043	0.288±0.086	2.1	0.898±0.127
GRB 000301C	28.00±0.30	0.130	0.00629	0.066±0.028	0.156±0.047	0.4	1.054±0.462
GRB 000418	23.80±0.20	0.080	0.00044	0.096±0.027	0.345±0.103	0.5	0.239±0.670

[a]Since the GRB position pierces through the Large Magellanic Cloud (Djorgovski et al. 1999d), we have added 0.13 mag extinction to the Galactic extinction quoted in Pian (2001). This assumes an average extinction through the LMC of $E(B - V) = 0.05$ (Dutra et al. 2001).

Note. — Column 2 gives the de-reddened host magnitude as referenced in Pian (2001), Djorgovski et al. (2001) and Sokolov et al. (2001). Column 3 gives the estimated extinction in the direction of the GRB host galaxy Pian (2001). Column 4 gives the estimated probability that the assigned host is a chance superposition and not physically related to the GRB (following §6.6.1). The half-light radii R_{half} are observed from HST imaging (col. 5) or calculated using the magnitude-radius empirical relationship (col. 6; see text). For HST imaging, the uncertainty is taken as the sum of the statistical error and estimated systematics error (0".025 which is approximately the size of one de-convolved pixel). Otherwise, the uncertainty is taken as 30% of the calculated radius (col. 6). Column 7 gives the estimated host disk scale length. The host-normalized offset $r_0 = R_0/R_{half}$ given in col. 8 is derived from (if possible) the observed half-light radius or the calculated half-light radius (otherwise). The error on r_0 is σ_r from equation 6.8. Note that $r_0 - \sigma_r \leq r \leq r_0 + \sigma_r$ is not necessarily the 68% percent confidence region of the true offset since the probability distribution is not Gaussian.

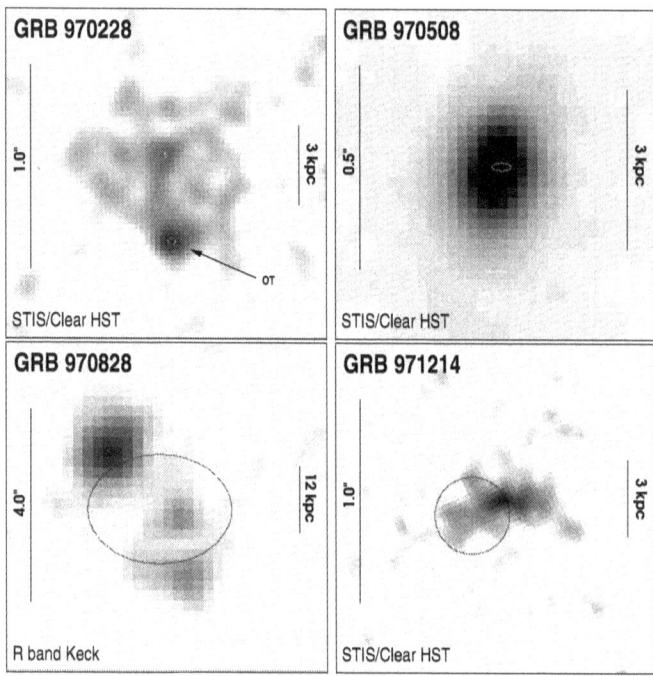

Figure 6.2 The location of individual GRBs about their host galaxies. The ellipse in each frame represents the 3-σ error contour for the location of the GRB as found in §6.5 and in table 6.2. The angular scale of each image is different and noted on the left-hand side. The scale and stretch was chosen to best show both the detailed morphology of the host galaxy and the spatial relationship of the GRB and the host. The GRB afterglow is still visible is some of the images (GRB 970228, GRB 991216). In GRB 980425, the location of the associated supernova is noted with an arrow. In all cases where a redshift is available for the host or GRB afterglow, we also provide a physical scale of the region on the right-hand side of each image. For clarity, the host centers are marked with "×" when the centers are not obvious. For all images, North is up and East is to the left.

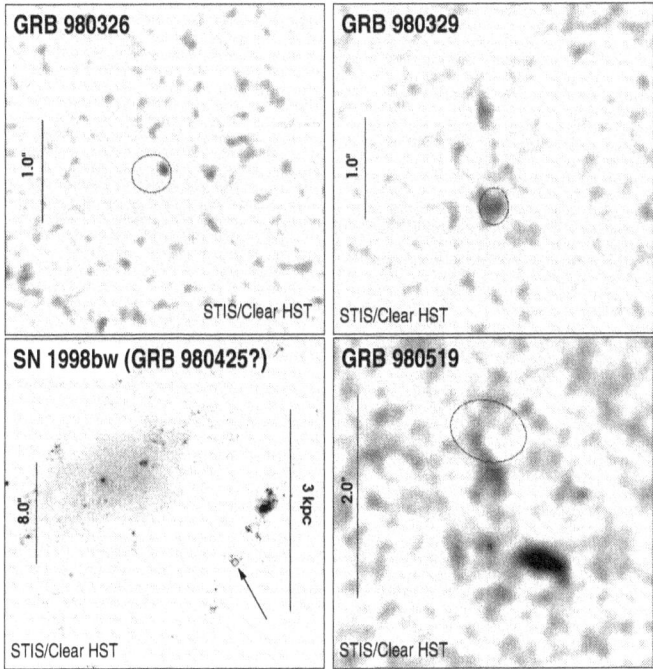

Figure 6.2 (cont.) The location of individual GRBs about their host galaxies.

6.5. INDIVIDUAL OFFSETS AND HOSTS

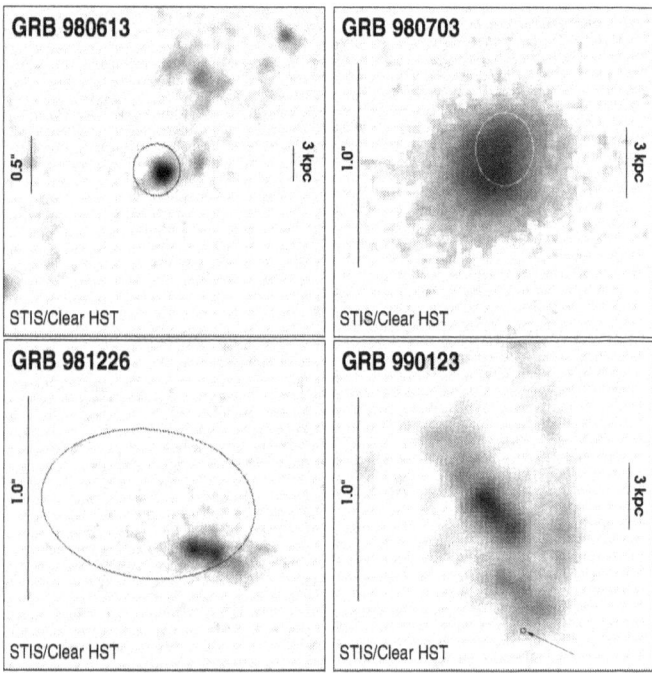

Figure 6.2 (cont.) The location of individual GRBs about their host galaxies.

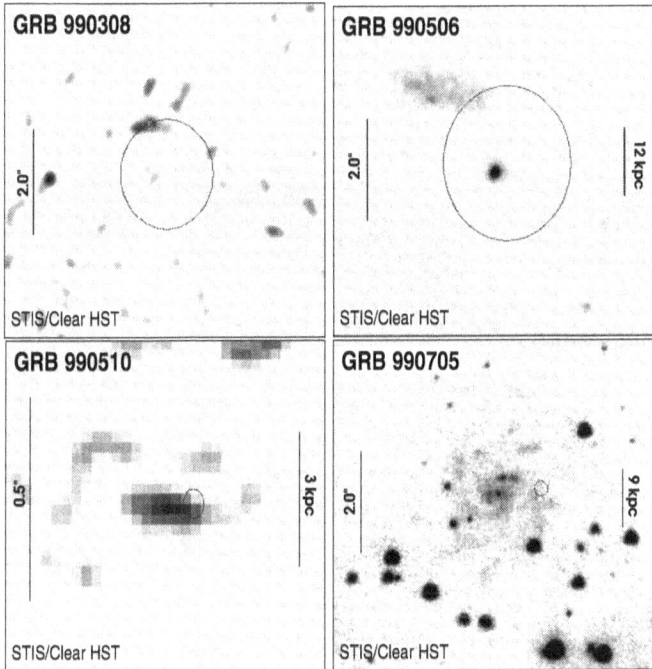

Figure 6.2 (cont.) The location of individual GRBs about their host galaxies.

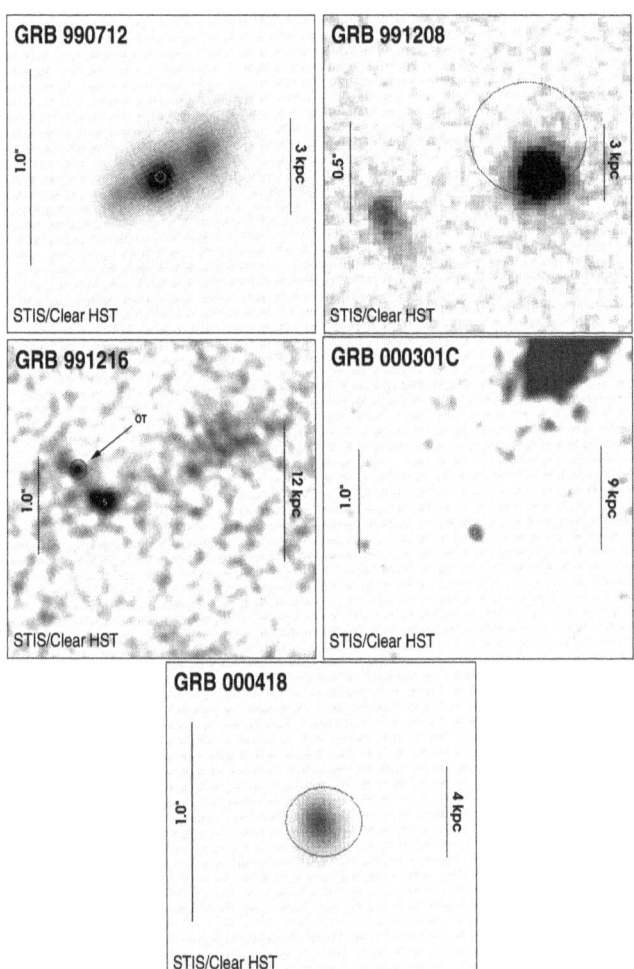

Figure 6.2 (cont.) The location of individual GRBs about their host galaxies.

of $R \approx 27.3$ mag. Given the close spatial connection of other GRBs with galaxies, we posited that a deeper integration would reveal a nearby host. Indeed, Fruchter et al. (2001a) recently reported the detection with HST/STIS imaging of a very faint ($V = 29.25 \pm 0.25$ mag) galaxy within 25 mas of the OT position.

For astrometry we used an R-band image, from 27 April 1998 when the OT was bright and found the position of the OT on our deep R-band image from 18 December 1998. In this deep R-band image we found 34 objects in common with the HST/STIS drizzled image. We confirm the presence of this faint and compact source near the OT position though our astrometry places the OT at a distance of 130 mas ($\sigma_r = 68$ mas).

The galaxy and OT position are shown in figure 6.2. The low-level flux to the Southeast corner of the image is a remnant from a diffraction spike of a nearby bright star. Fruchter et al. (2001a) find that the putative host galaxy is detected at the 4.5-σ level. Adding to the notion that the source is not some chance superposition, we note that the galaxy is the brightest object within 3×3 arcsec^2 of the GRB position. There is also a possible detection of a low-surface brightness galaxy $\sim$0.5 arcsec to the East of the galaxy.

6.5.6 GRB 980329

The afterglow of GRB 980329 was first detected at radio wavelengths (Taylor et al. 1998). Our best early time position was obtained using Keck K-band image of the field observed by J. Larkin and collaborators (Larkin et al. 1998). We recently obtained deep R- imaging of the field with Keck/ESI and detected the host galaxy at $R = 26.53 \pm 0.22$ mag. We found the location of the afterglow relative to the host using 13 stars in common to the early K-band and late R-band image.

As shown in figure 6.2, the GRB is coincident with a slightly extended faint galaxy. Our determined angular offset (see table 6.2) of the GRB from this galaxy is significantly closer to the putative host than the offset determined by Holland et al. (2000f) in late-time HST imaging (our astrometric uncertainties are also a factor of ~ 9 smaller). The difference is possibly explained by noting that the Holland et al. (2000f) analysis used the VLA radio position and just three USNO-A2.0 stars to tie the GRB position to the HST image.

6.5.7 GRB 980425

The SN 1998bw was well-localized at radio wavelengths (Kulkarni et al. 1998) with an astrometric position relative to the ICRS of 100 mas in each coordinate. Ideally, we could calibrate the HST/STIS image to ICRS to ascertain where the radio source lies. However, without Hipparcos/Tycho astrometric sources or radio point sources in the STIS field, such absolute astrometric positioning is difficult.

Instead, we registered an early ground-based image to the STIS field to determine the differential astrometry of the optical SN with respect to its host. Unfortunately, most early images were relatively shallow exposures to avoid saturation of the bright SN and so many of the point sources in the STIS field are undetected. The best seeing and deepest exposure from ground-based imaging is from the EMMI/ESO NTT 3.5 meter Telescope on 4.41 May 1998 (Galama 1999) where the seeing was 0.9 arcsec FWHM. We found 6 point sources which were detected in both the STIS/CLEAR and the ESO NTT I-band image. The use of I-band positions for image registration is justified since all 6 point sources are red in appearance and therefore unlikely to introduce a systematic error in the relative positioning. Since the number of astrometric tie sources is low, we did not fit for high-order distortions in the ESO image and instead we fit for the relative scale in both the x and y directions, rotation, and shift (5 parameters for 12 data points). We compute an r.m.s. uncertainty of 40 mas and 32 mas in the x and y positions of the astrometric tie sources. These transformation uncertainties dominate the error in the positional uncertainty of the SN in the ESO NTT image and so we take the transformation uncertainties as the uncertainty in the true position of the supernova with respect to the STIS host image.

The astrometric mapping places the optical position of SN 1998bw within an apparent star-forming region in the outer spiral arm of the host 2.4 kpc in projection at $z = 0.0088$ to the south-west of the galactic nucleus. Within the uncertainties of the astrometry the SN is positionally coincident with a bright, blue knot within this region, probably an HII region. This is consistent with the independent astrometric solutions reported by Fynbo et al. (2000).

6.5.8 GRB 980519

The GRB afterglow was well-detected in our early-time image from the Palomar 200-inch. We found 150 objects in common to this image and our intermediate-time Keck image. An astrometric registration between the two epochs was performed using IRAF/GEOMAP. Based on this astrometry, Bloom et al. (1998d) reported the OT to be astrometrically consistent with a faint galaxy, the putative host. This is the second faintest host galaxy (after GRB 990510; see below) observed to date with $R = 26.1 \pm 0.3$.

We found 25 objects in common with the intermediate-time Keck image and the HST/STIS image. These tie objects were used to further propagate the OT position onto the HST frame. Inspection of our final HST image near the optical transient location reveals the presence of low surface-brightness emission connecting the two bright elongated structures. Morphologically, the "host" appears to be tidally interacting galaxies, although this interpretation is subjective. The GRB location is coincident with the dimmer elongated structure to the north. Using the approximate geometric center of the host, we estimate the center, albeit somewhat arbitrarily, as the faint knot south of the GRB location and $\sim 0''.3$ to the east of the brighter elongated structure. The half-light radius of the system was also measured from this point. From this "center" we find the offset of the GRB given in table 6.2.

6.5.9 GRB 980613

The morphology of the system surrounding the GRB is complex and discussed in detail in Djorgovski et al. (2000). There we found the OT to be within $\sim$3 arcsec of five apparent galaxies or galaxy fragments, two of which are very red ($R - K > 5$). In more recent HST imaging, the OT appears nearly coincident with a compact high-surface brightness feature, which we now identify as the host center. Given the complex morphology, we chose to isolate the feature in the determination of the half-light radius by truncating the curve-of-growth analysis at 0.5 arcsec from the determined center.

6.5.10 GRB 980703

The optical transient was well-detected in our early time image and, based on the light curve and the late-time image, the light was not contaminated by light from the host galaxy. Berger et al. (2001a) recently found that the radio transient was very near the center of the radio emission from the host.

We found 23 objects in common to the Keck image and our final reduced HST/STIS image and computed the geometric transformation. The r.m.s. uncertainty of the OT position on the HST image was quite small: 49 mas and 60 mas in the instrumental x and y coordinates, respectively. We determined the center of the host using IRAF/ELLIPSE and IRAF/CENTER which gave consistent answers to 2 mas in each coordinate.

Recently, Berger et al. (2001a) compared the VLBA position of the afterglow with the position of the persistent radio emission from the host. Since both measurements were referenced directly to the ICRS, the offset determined was a factor of $\sim$3 times more accurate than that found using the optical afterglow; the two offset measurements are consistent within the errors. In the interest of uniformity, we use the optical offset measurement in the following analysis.

6.5.11 GRB 981226

Unfortunately, no optical transient was found for this burst though a radio transient was identified (Frail et al. 1999). We rely on the transformation between the USNO-A2.0 and the Keck image to place the host galaxy position on the ICRS (see Frail et al. 1999, for further details). We then determine the location of the radio transient in the HST frame using 25 compact sources common to both the HST and Keck image. In figure 6.1 we show as example the tie objects in both the Keck and HST image. The tie between the two images is excellent with an r.m.s. uncertainty of 33 mas and 47 mas in the instrumental x and y positions. Clearly, the uncertainty in the radio position on the Keck image dominates the overall location of the GRB on the HST image.

The host appears to have a double nucleated morphology, perhaps indicative of a merger or interacting system. Hjorth et al. (2000) noted, by inspecting both the STIS Longpass and the STIS clear image, that the north-eastern part of the galaxy appeared significantly bluer then the south-western part. As expected from these colors, the center of the host, as measured in our late-time R-band Keck, lies near ($\sim$ 50 mas) the centroid of the red (south-western) portion of the host. We assign the R-band centroid in Keck image as the center of the host.

6.5.12 GRB 990123

This GRB had an extremely bright prompt optical afterglow emission which was found in archived images from a robotic telescope, the Robotic Optical Transient Search Experiment (Akerlof et al. 1999). We reported on the astrometric comparison of ground-based data with HST imaging and found that the bright point source on the southern edge of a complex morphological system was the afterglow (Bloom et al. 1999b). Later HST imaging revealed that indeed this source did fade (e.g., Fruchter et al. 1999) as expected of GRB afterglow.

As seen in figure 6.2, the host galaxy is fairly complex, with two bright elongated regions spaced by $\sim 0''.5$ which run approximately parallel to each other. The appearance of spatially curved emission to the west may be a tidal tail from the merger of two separate systems or a pronounced spiral arm of the brighter elongated region to the north. We choose, again somewhat subjectively, the peak of this brighter region as the center of the system and find the astrometric position of the GRB directly from the first HST epoch.

6.5.13 GRB 990308

An optical transient associated with GRB 990308 was found by Schaefer et al. (1999). Though the transient was detected at only one epoch (3.3 hours after the GRB; Schaefer et al. 1999), it was observed in three band-passes, twice in R-band. Later-time Keck imaging revealed no obvious source at the location of the transient to $R = 25.7$ mag, suggesting that the source had faded by at least $\sim$7.5 mag in R-band.

A deep HST exposure of the field was obtained by Holland et al. (2000e) who reported that the Schaefer et al. position derived from the USNO-A2.0 was consistent with two faint galaxies.

We found the offset by two means. First, we found an absolute astrometric solution using 12 USNO-A2.0 stars in common with the later-time Keck image. The HST/STIS and the Keck R-band image were then registered using 27 objects in common. Second, we found a differential position by using early ground-based images kindly provided by B. Schaefer to tie the optical afterglow position directly to the Keck (then to HST) image. Both methods give consistent results though the differential method is, as expected, more accurate.

Our astrometry places the OT position further East from the two faint galaxies than the position derived by Holland et al. (2000e). At a distance of 0.73 arcsec to the North of our OT position, there appears to be a low-surface brightness galaxy near the detection limit of the STIS image (see figure 6.2), similar to the host of GRB 980519 (there is also, possibly, a very faint source 0.23 arcsec southwest of the OT position, but the reality of its detection is questionable). Due to the faintness and morphological nature of the source, a detection confidence limit is difficult to quantify, but we are reasonably convinced that the source is real. At $V \sim 27$ mag, the non-detection of this galaxy in previous imaging is consistent with the current STIS detection. Since the angular extent of the galaxy spans $\sim$25 drizzled STIS pixels

(∼0.63 arcsec), more high-resolution HST imaging is not particularly useful for confirming the detection of the galaxy. Instead, deeper ground-based imaging with a large aperture telescope would be more useful.

6.5.14 GRB 990506

The Keck astrometric comparison to the radio position was given in Taylor et al. (2000), with a statistical error of 250 mas. We transferred this astrometric tie to the HST/STIS image using 8 compact sources common to both the Keck and HST images of the field near GRB 990506. The resulting uncertainty is negligible compared to the uncertainties in the radio position on the Keck image. As first reported in Taylor et al. (2000), the GRB location appears consistent with a faint compact galaxy. Holland et al. (2000d) later reported that the galaxy appears compact even in the STIS imaging.

6.5.15 GRB 990510

This GRB is well-known for having exhibited the first clear evidence of a jet manifested as an achromatic break in the light curve (e.g., Harrison et al. 1999; Stanek et al. 1999). Recently, we discovered the host galaxy in late-time HST/STIS imaging (Bloom 2000) with $V = 28.5 \pm 0.5$. Registration of the early epoch where the OT was bright reveals the OT occurred 64 ± 9 mas west and 15 ± 12 mas north of the center of the host galaxy. This amounts to a significant displacement of 66 ± 9 mas or 600 pc at a distance of $z = 1.62$ (Galama et al. 1999). The galaxy is extended with a position angle PA = 80.5 ± 1.5 degree (east of north) and an ellipticity of about ∼0.5.

In retrospect, the host does appear to be marginally detected in the July 1999 imaging as well as the later April 2000 image although, at the time, no galaxy was believed to have been detected (Fruchter et al. 1999).

6.5.16 GRB 990705

Masetti et al. (2000) discovered the infrared afterglow of GRB 990705 projected on the outskirts of the Large Magellanic Cloud. At the position of the afterglow, Masetti et al. (2000) noted an extended galaxy seen in ground-based V-band imaging; they identified this galaxy as the host. Holland et al. (2000a) reported on HST imaging of the field and noted, thanks to the large size (∼ 2 arcsec) of the galaxy and resolution afforded by HST, an apparent face-on spiral at the location of the transient. We retrieved the public HST data and compared the early images provided by N. Masetti with our final reduced HST image. Consistent with the position derived by Holland et al. (2000a), we find that the transient was situated on the outskirts of a spiral arm to the west of the galaxy nucleus and just north of an apparent star-forming region.

6.5.17 GRB 990712

This GRB is the lowest measured redshift of a "cosmological" GRB with $z = 0.4337$ (Hjorth et al. 2000). Unfortunately, the astrometric location of the GRB appears to be controversial, though there is no question that the GRB occurred within the bright galaxy pictured in figure 6.2. Hjorth et al. (2000) found that the only source consistent with a point source in the earlier HST image was the faint region to the Northwest side of the galaxy and concluded that the source was the optical transient. However, Fruchter et al. (2000c) found that this source did not fade significantly. Instead the Fruchter et al. analysis showed, by subtraction of two HST epochs, that a source did fade near the bright region to the southeast. While the fading could be due to AGN activity instead of the presence of a GRB afterglow, we adopt the conclusion of Fruchter et al. for astrometry and place conservative uncertainties on the location relative to the center as 75 mas (3 pixels) in both α and δ for 1-σ errors. We did not conduct an independent analysis to determine this GRB offset.

6.5.18 GRB 991208

In our early K-band image of the field, we detect the afterglow as well as 7 suitable tie stars to our late ESI image. The host galaxy is visible in the ESI image and the subsequent offset was reported by Diercks et al. (2000). An HST image was later obtained and reported by Fruchter et al. (2000d) confirming the presences of the host galaxy.

We reduced the public HST/STIS data on this burst and found the offset in the usual manner by tying the OT position from Keck to the HST frame. The GRB afterglow position falls near a small, compact galaxy. A fainter galaxy, to the Southeast, may also be related to the GRB/host galaxy system (see figure 6.2).

6.5.19 GRB 991216

We used nine compact objects in common to our early Keck image (seeing FWHM = $0''.66$) and the late-time HST/STIS imaging to locate the transient. As noted first by Vreeswijk et al. (2000), the OT is spatially coincident with a faint, apparent point source in the HST/STIS image. Our astrometric accuracy of $\sigma_r = 32$ mas of the OT position is about four times better than that of Vreeswijk et al. (2000). Thanks to this we can confidently state that the OT coincides with a point source on the HST/STIS image. We believe this point source, as first suggested by Vreeswijk et al., is the OT itself.

The "location" of the host galaxy is difficult to determine. The OT does appear to reside to the southwest of faint extended emission (object "N" from Vreeswijk et al. 2000) but it is also located to the northeast of a brighter extended component (object "S" from Vreeswijk et al. 2000). There appears to be a faint bridge of emission connecting the two regions as well as the much larger region to the west of the OT (see figure 6.2). In fact, these three regions may together comprise a large, low-surface brightness system. Again, somewhat arbitrarily, we take the center of the "host" to be the peak of object "S."

6.5.20 GRB 000301C

Fruchter et al. (2000a), in intermediate-time (April 2000) imaging of the field of GRB 000301C, detected a faint unresolved source coincident with the location of the GRB afterglow; the authors reckoned the source to be the faded afterglow itself. In the most recent imaging on February 2001 the same group detected a somewhat fainter, compact object very near the position of the transient. Given the time ($\sim$ one year) since the GRB, the authors suggested that the afterglow should have faded below the detection level and that therefore this object is the host of GRB 000301C (Fruchter et al. 2001b).

We confirm the detection of this source and measure the offset using earlier time imaging from HST. Though no emission line redshift of this source has been obtained given its proximity to the GRB, it likely resides at $z = 2.03$, inferred from absorption spectroscopy (Smette et al. 2001; Castro et al. 2000) of the OT.

In figure 6.2 we present the late-time image from HST/STIS. A galaxy $2''.13$ from the transient to the northwest is detected at $R = 24.25 \pm 0.08$ mag and may be involved in possible microlensing of the GRB afterglow (e.g., Garnavich et al. 2000).

6.5.21 GRB 000418

We reported the detection of an optically bright component and an infrared bright component at the location of GRB 000418 (Bloom et al. 2000b). Metzger et al. (2000) later reported that HST/STIS imaging of the field revealed that the OT location was $0''.08 \pm 0''.15$ east of the center of the optically bright component, a compact galaxy.

For our astrometry we used an early Keck R-band image and the late HST/STIS image. The astrometric uncertainty is improved over the Metzger et al. (2000) analysis by a factor of 2.4. Within errors, the OT is consistent with the center of the compact host.

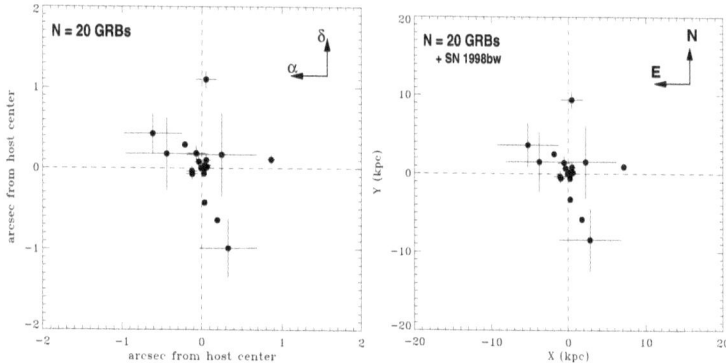

Figure 6.3 (left) The angular distribution of 20 gamma-ray bursts about their presumed host galaxy. The error bars are 1 σ and reflect the total uncertainty in the relative location of the GRB and the apparent host center. The offset of GRB 980425 from its host is suppressed for clarity since the redshift, relative to all the others, GRB was so small. (right) The projected physical offset distribution of 20 γ-ray bursts (now including SN1998bw/GRB 980425) about their presumed host galaxies. The physical offset is assigned assuming $H_0 = 65$ km/s Mpc^{-1}, $\Lambda = 0.7$, and $\Omega_m = 0.3$ and assuming the GRB and the presumed host are at the same redshift. Where no redshift has been directly measured a redshift is assigned equal to the median redshift ($z = 0.966$) of all GRBs with measured redshifts (see text).

SECTION 6.6

The Observed Offset Distribution

6.6.1 Angular offset

As seen in table 6.2 and §6.5, there are 20 GRBs for which we have a reliable offset measurement from self-HST, HST→HST, HST→GB, GB→GB, or RADIO→OPT astrometric ties. There are several representations of this data worth exploring. In figure 6.3 we plot the angular distribution of GRBs about their presumed host galaxy. In this figure and in the subsequent analysis we exclude GRB 980425 because the association of this GRB with SN 1998bw is still controversial. And more importantly (for the purposes of this paper) the relation of GRB 980425 with the classical "cosmological" GRBs is unclear (Schmidt 1999) given that, if the association proved true, the burst would have been under-luminous by a factor of $\sim 10^5$ (Galama et al. 1998b; Bloom et al. 1998a).

As can be seen from table 6.2 and figure 6.3, well-localized GRBs appear on the sky close to galaxies. The median projected offset of the 20 GRBs from their putative host galaxies is 0.17 arcsecond— sufficiently small that almost all of the identified galaxies must be genuine hosts (see below). In detail, three of the bursts show no measurable offset from the centroid of their compact hosts (970508, 980703, 000418) whereas five bursts appear well displaced ($\gtrsim 0''.3$) from the center of their host at a high level of significance. Three additional bursts detected via radio afterglows (GRB 970828, GRB 981226, GRB 990506) and GRB 990308 (poor astrometry of the discovery image due to large pixels and shallow depth) suffer from larger uncertainties (r.m.s. ≈ 0.3 arcsecond) but have plausible host galaxies.

As discussed in Appendix 6.A, GB→GB or GB→HST astrometry could systematically suffer from the effects of differential chromatic refraction, albeit on the 5–10 mas level. The HST→HST measured offsets of GRB 970228, GRB 970508, GRB 990123, GRB 990510, GRB 990712, and GRB 000301C are immune from DCR effects. Since optical transients are, in general, red in appearance and their hosts

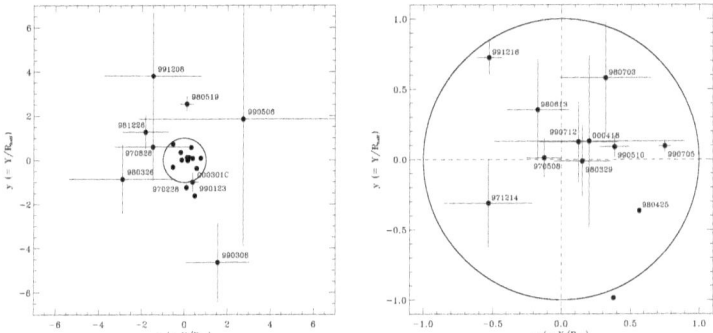

Figure 6.4 Host-normalized offset distribution. The dimensionless offsets are the observed offsets (X_0, Y_0) normalized by the host half-light radius (R_{half}) of the presumed host galaxy. See text for an explanation of how the half-light radius is found. The 1-σ error bars reflect the uncertainties in the offset measurement and in the half-light radius. As expected if GRBs occur where stars are formed, there are 10 GRBs (plus 1998bw/GRB 980425) inside and 10 GRBs outside the half-light radius of their host. (left) All GRBs outside of one half-light radius (small circle) are labeled. (right) All GRBs observed to be internal to one half-light radius are labeled.

blue, DCR will systematically appear to pull OTs away from their hosts in the parallactic direction toward the horizon. Comparing the observed offsets directions parallactic at the time of each OT observation in table 6.2, we find no systematic correlation thus confirming that DCR does not appear to play a dominant role in determining the differential offsets of OTs from their hosts.

On what basis can we be confident that the host assignment is correct for a particular GRB? Stated more clearly in the negative is the following question: "What is the probability of finding an unrelated galaxy (or galaxies) within the localization error circle of the afterglow (3-sigma) or, in the case where the localization error circle is very small, whether a galaxy found close to a GRB localization is an unrelated galaxy seen in projection?" This probability, assuming that the surface distribution of galaxies is uniform and thus follows a Poisson distribution (i.e., we ignore clustering of galaxies) is

$$P_{i,\text{chance}} = 1 - \exp(-\eta_i). \qquad (6.1)$$

Here

$$\eta_i = \pi r_i^2 \sigma(\leq m_i) \qquad (6.2)$$

is the expected number of galaxies in a circle with an effective radius, r_i, and

$$\sigma(\leq m_i) = \frac{1}{3600^2 \times 0.334 \log_e 10} \times 10^{0.334(m_i - 22.963) + 4.320} \text{ galaxy arcsec}^{-2}$$

is the mean surface density of galaxies brighter than R-band magnitude of m_i, found using the results from Hogg et al. (1997). Since GRB are observed through some Galactic extinction, the surface density of galaxies at a given limiting flux is reduced; therefore, we use the reddened host galaxy magnitude for m_i (= col. 2 − col. 3 of table 6.3).

There are few possible scenarios for determining r_i at a given magnitude limit. If the GRB is

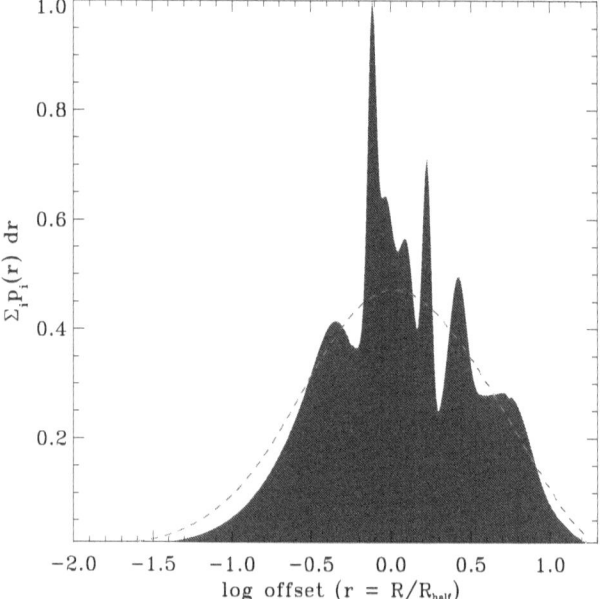

Figure 6.5 The GRB offset distribution as a function of normalized galactocentric radius. The normalized offset is $r = R/R_{\rm half}$, where R is the projected galactocentric offset of the GRB from the host and $R_{\rm half}$ is the half-light radius of the host. This distribution is essentially a smooth histogram of the data, but one which takes into account the uncertainties in the measurements: the sharper peaks are due to individual offsets where the significance (r_0/σ_{r_0}) of the offset is high. That is, if a GRB offset is well-determined, its contribution to the distribution will appear as a δ-function centered at $r = r_0$. The dashed curve is the distribution under the blue (dark) curve but smoothed with a Gaussian of FWHM = 0.7 dex in r. Strikingly, the peak of the probability is near one half-light radius, a qualitative argument for the association of GRBs with massive star formation. We compare in detail this distribution with predicted progenitor distributions in §6.7.

very well localized inside the detectable light of a galaxy, then $r_i \approx 2 R_{\text{half}}$ is a reasonable estimate to the effective radius. If the localization is poor and there is a galaxy inside the uncertainty position, then $r_i \approx 3\sigma_{R_0}$. If the localization is good, but the position is outside the light of the nearest galaxy, then $r_i \approx \sqrt{R_0^2 + 4 R_{\text{half}}^2}$. Therefore, we take $r_i = \max[2 R_{\text{half}}, 3\sigma_{R_0}, \sqrt{R_0^2 + 4 R_{\text{half}}^2}]$ as a conservative estimate to the effective radius. Here, the quantity R_0 is the radial separation between the GRB and the presumed host galaxy, R_{half} is the half-light radius, and σ_{R_0} is the associated r.m.s. error (see table 6.2).

If no "obvious" host is found (i.e., $P_{\text{chance}} \gtrsim 0.1$) then we often seek deeper imaging observations, which will, in general, decrease the estimated r_i as more and more galaxies are detected. However, the estimate for η_i should remain reasonable since the surface density of background galaxies continues to grow larger with increasing depth. This is to say that there is little penalty to pay in statistically relating sky positions to galaxies by observing to fainter depths.

The values for $P_{i,\text{chance}}$ are computed and presented in table 6.3. As expected, GRBs which fall very close to a galaxy (e.g., GRB 970508, GRB 980703, GRB 990712) are likely to be related to that galaxy. Similarly, GRB localizations with poor astrometric accuracy (e.g., GRB 990308, 970828) yield larger probabilities that the assigned galaxy is unrelated.

In the past, most authors (including ourselves) did not endeavor to produce a probability of chance association, instead opting to assume that these assigned galaxies are indeed the hosts. Nevertheless, we believe these estimates are conservative; for instance, van Paradijs et al. (1997) estimated that P_{chance} (970228) = 0.0016 which is a value 5.8 times smaller than our estimate. Again, we emphasize that the estimated probabilities are constructed *a posteriori* so there is no exact formula to the determine the true P_{chance}.

The probability that **all** supposed host galaxies in our sample are random background galaxies is

$$P(n_{\text{chance}} = m = \text{all}) = \prod_{k=1}^{m} P_k,$$

with $m = 20$ and P_k found from equation 6.1 for each GRB k. Not surprisingly, this number is extremely small, $P(\text{all}) = 2 \times 10^{-60}$, insuring that at least some host assignments must be correct.

The probability that **all** galaxies are physically associated (i.e., that none are chance super-position of a random field galaxies) is

$$P(n_{\text{chance}} = 0) = \prod_{k=1}^{m}(1 - P_k) = 0.483.$$

In general, the chance that n assignments will be spurious out of a sample of $m \geq n$ assignments is

$$P(n_{\text{chance}}) = \frac{1}{n_{\text{chance}}!} \times \qquad (6.3)$$

$$\overbrace{\sum_{i}^{m}\sum_{j \neq i}^{m} \cdots}^{n_{\text{chance}}} \left[\overbrace{P_i \times P_j \times \cdots}^{n_{\text{chance}}} \prod_{k \neq i \neq j \neq \cdots}^{m}(1 - P_k) \right].$$

$P(n_{\text{chance}})$ reflects the probability that we have generated a number n_{chance} of spurious host galaxy identifications. For our sample, we find that $P(1) = 0.305$, $P(2) = 0.106$, and $P(3) = 0.015$ and so the number of spurious identifications is likely to be small, ~ 1–2. Indeed, if the two GRBs with the largest P_{chance} are excluded (GRB 970828, GRB 990308), then $P(n_{\text{chance}} = 0)$ jumps to 0.76. Thus we are confident that almost all of our identifications are quite secure.

The certainty of our host assignments of the nearest galaxy to a GRB finds added strength by using redshift information. In *all* cases where an absorption redshift is found in a GRB afterglow (GRB 970508,

GRB 980613, GRB 990123, GRB 980703, GRB 990712, GRB 991216), the highest redshift absorption system is observed to be at nearly the same emission redshift of the nearest galaxy. Therefore, with these bursts, clearly the nearest galaxy cannot reside at a higher redshift than the GRB. The galaxy may simply be a foreground object which gives rise both to nebular line emission and the absorption of the afterglow originating from a higher redshift. However, using the observed number density evolution of absorbing systems, Bloom et al. (1997) calculated that statistically in $\gtrsim 80\%$ of such absorption cases, the GRB could reside no further than 1.25 times the absorption redshift. For example, if an emission/absorption system is found at $z = 1.0$, then there is only a $\lesssim 20\%$ chance that the GRB could have occurred beyond redshift $z = 1.25$ without another absorption system intervening. Though this argument cannot prove that a given GRB progenitor originated from the assigned host, the effect of absorption/emission redshifts is to confine the possible GRB redshifts to a shell in redshift-space, reducing the number of galaxies that could possibly host the GRB, and increasing the chance that the host assignment is correct. Therefore, given this argument and the statistical formulation above, we proceed with the hypothesis that, as a group, GRBs are indeed physically associated with galaxies assigned as hosts.

6.6.2 Physical projection

Of the 20 GRBs with angular offsets, five have no confirmed redshift, and the angular offset is thus without a physical scale. These bursts have hosts fainter than $R \approx 25$ mag and, given the distribution of other GRB redshifts with these host magnitudes, it is reasonable to suppose that the five bursts originated somewhere in the redshift range $z = 0.5$–5. It is interesting to note (with our assumed cosmology) that despite a luminosity distance ratio of 37 between these two redshifts, the angular scales are about the same: $D_\theta(z = 0.5)/D_\theta(z = 5) \approx 1$. In fact, over this entire redshift range, 6.6 kpc arcsec^{-1} < $D_\theta(z)$ < 9.1 kpc arcsec^{-1} which renders the conversion of angular displacement to physical projection relatively insensitive to redshift. For these five bursts, then, we assign the median D_θ of the other bursts with known redshifts so that $D_\theta = 8.552$ kpc arcsec^{-1} (corresponding to a redshift of $z = 0.966$) and scale the observed offset uncertainty by an additional 30%. Here, we use the GRB redshifts (and, below, host magnitudes) compiled in the review by Kulkarni et al. (2000). The resulting physical projected distribution is depicted in figure 6.3 and given in table 6.2. The median projected physical offset of the 20 GRBs in the sample is 1.31 kpc or 1.10 kpc including only those 15 GRBs for which a redshift was measured. The minimum offset found is just 91 ± 90 pc from the host center (GRB 970508).

6.6.3 Host-normalized projected offset

If GRBs were to arise from massive stars, we would then expect that the distribution of GRB offsets would follow the distribution of the light of their hosts. As can be seen in figure 6.2, qualitatively this appears to be the case since almost all localizations fall on or near the detectable light of a galaxy.

The next step in the analysis is to study the offsets but normalized by the half-light radius of the host. This step then allows us to consider all the offsets in a uniform manner. The half-light radius, R_{half}, is estimated directly from STIS images with sufficiently high signal-to-noise ratio and in the remaining cases we use the empirical half-light radius–magnitude relation of Odewahn et al. (1996b); we use the de-reddened R-band magnitudes found in the GRB host summaries from Djorgovski et al. (2001) and Sokolov et al. (2001). Table 6.3 shows the angular offsets and the effective radius used for scaling. Where the empirical half-light radius–magnitude relation is used, we assign an uncertainty of 30% to R_{half}.

The median of the distribution of normalized offsets is 0.976 (table 6.3). That this number is close to unity suggests a strong correlation of GRB locations with the light of the host galaxies. The same strong correlation can be graphically seen in figure 6.4 where we find that half of the galaxies lie inside the half-light radius and the remaining, outside the half-light radius. We remark that the effective wavelength of the STIS band-pass and the ground-based R band correspond to rest-frame UV and thus GRBs appear to be traced quite faithfully by the UV light which mainly arises from the youngest and thus massive stars. We will examine the distribution in the context of massive star progenitors more closely in §6.7.2.

6.6.4 Accounting for the uncertainties in the offset measurements

A simple way to compare the normalized offsets to the expectations of various progenitor models (see §6.7) is through the histogram of the offsets. However, due to the small number of offsets, the usual binned histogram is not very informative. In addition, the binned histogram implicitly assumes that the observables can be represented by δ-functions and this is not appropriate for our case, in which several offsets are comparable to the measurement uncertainty.

To this end we have developed a method to construct a probability histogram (PH) that takes into account the errors on the measurements. Simply put, we treat each measurement as a probability distribution of offset (rather than a δ-function) and create a smooth histogram by summing over all GRB probability distributions. Specifically, for each offset i we create an individual PH distribution function, $p_i(r)\,dr$, representing the probability of observing a host-normalized offset r for that burst. The integral of $p_i(r)\,dr$ is normalized to unity. The total PH is then constructed as $p(r)\,dr = \sum_i p_i(r)\,dr$ and plotted as a shaded region curve in figure 6.5; see Appendix 6.B for further details.

The total cumulative probability histogram, $\int_0^r p(r)\,dr$, is depicted as the solid smooth curve in figure 6.6. There is, as expected, a qualitative similarity between the cumulative total PH distribution and the usual cumulative histogram distribution.

SECTION 6.7

Testing Progenitor Model Predictions

Given the observed offset distribution, we are now in the position to pose the question: which progenitor models are favored by the data? Clearly, GRBs as a class do not appear to reside at the centers of galaxies, and so we can essentially rule out the possibility that *all* GRBs localized to-date arise from nuclear activity.

6.7.1 Delayed merging remnants binaries (BH–NS and NS–NS)

In general, the expected distributions of merging remnant binaries are found using population synthesis models for high-mass binary evolution to generate synthetic remnant binaries. The production rate of such binaries from other channels (such as three-body encounters in dense stellar clusters) are assumed to be small relative to isolated binary evolution. Due to gravitational energy loss, the binary members eventually coalesce but may travel far from their birth-site before doing so. The locations of coalescence are determined by integrating the synthetic binary orbits in galactic potential models.

Bloom et al. (1999a), Fryer et al. (1999a) and Bulik et al. (1999) have simulated the expected radial distribution of GRBs in this manner. All three studies essentially agree on the NS–NS differential offset distributions as a function of host galaxy mass[3] The NS–NS distributions of Bloom et al. (1999a) are slightly more concentrated toward smaller offset radii than those from Fryer et al. (1999a) and Bulik et al. (1999); To account for this, Fryer et al. (1999a) suggested that the Bloom et al. (1999a) synthesis may have incorrectly predicted an over-abundance of compact binaries with small merger ages, because the population synthesis did not include a non-zero helium star radius; this is not the case, although an arithmetic error in our code may account for the discrepancy (Sigurdsson, priv. communication). We

---

[3] It appears that figure 22 of Fryer et al. (1999a), the cumulative distribution of merger sites, is mislabeled (showing NS–NS merger sites to be at radii a factor of ∼10 times larger than suggested in their differential distributions and tables). Accepting the radii given in figure 21 and table 10 of Fryer et al. (1999a), all three of the aforementioned studies are in approximate agreement on the NS–NS merger sites.

emphasize that, within a factor of a ~two, population synthesis studies that use the classical channels for NS–NS production, are in agreement with NS–NS merger sites with respect to host galaxies.

The formation scenarios of BH–NS binaries are less certain than that of NS–NS binaries. Both Fryer et al. (1999a) and Belczyński et al. (2000) suggest that so-called "hypercritical accretion" (Bethe & Brown 1998) dominates the birthrate of BH–NS binaries. Briefly, hypercritical accretion occurs when the primary star evolves off the main sequence and explodes as a supernova, leaving behind a neutron star. Mass is rapidly accreted from the secondary star (in red giant phase) during common envelope evolution, causing the primary neutron star to collapse to a black hole. The secondary then undergoes a supernova explosion leaving behind a NS. As in NS–NS binary formation, only some BH–NS systems will remain bound after having received systemic velocity kicks from two supernovae explosions.

One important difference is that BH–NS binaries are in general more massive (total system mass $M_{tot} \approx 5 M_\odot$) than NS–NS binaries ($M_{tot} \approx 3 M_\odot$). Furthermore, the coalescence timescale after the second supernova is shorter than in NS–NS binaries because of the BH mass. Therefore, despite similar evolutionary tracks, BH–NS binaries could be retained more tightly to host galaxies than NS–NS binaries (Bloom et al. 1999a; Belczyński et al. 2000). Belczyński et al. (2000) quantified this expected trend, showing that on average, BH–NS binaries merge ~few times closer to galaxies than NS–NS binaries. Surprisingly, Fryer et al. (1999a) found that BH–NS binaries merged *further* from galaxies than NS–NS binaries, but this result was not explained by Fryer et al. (1999a). Nevertheless, just as with NS–NS binaries, a substantial fraction of BH–NS binaries will escape the potential well of the host galaxy and merge well-outside of the host. For example, even in massive galaxies such as the Milky Way, these studies show that roughly 25% of mergers occur > 100 kpc from the center of a host galaxy.

Before comparing in detail the predicted and observed distributions, it is illustrative to note that the observed distribution appears qualitatively inconsistent with the delayed merging remnant binaries. All the population synthesis studies mentioned thus far find that at approximately 50% of merging remnants will occur outside of $\approx$ 10 kpc when the mass of the host is less than or comparable to the mass of the Milky Way. Comparing this expectation with figure 6.3, where no bursts lie beyond 10 kpc from their host, the simplistic Poisson probability that the observed distribution is the same as the predicted distribution is no larger than 2×10^{-3}.

To provide a more quantitative comparison of the observed distribution with the merging remnant expectation, we require a model of the location probability of GRB mergers about their hosts. These models, which should in principle vary from host to host, have a complex dependence on the population synthesis inputs, the location of star formation within the galaxies and the dark-matter halo mass.

No dynamical or photometric mass of a GRB host has been reported to-date. However, since many GRB hosts are blue starbursts (e.g., Djorgovski et al. 2001), it is not unreasonable to suspect that their masses will lie in the range of $0.001 - 0.1 \times 10^{11} M_\odot$ (e.g., Ostlin et al. 2001). The most obvious exceptions to this are the hosts of GRB 971214 and GRB 990705 which are likely to be near L_*. The observed median effective disk scale length of GRB hosts is $r_e = 1.1$ kpc though GRB hosts clearly show a diversity of sizes (table 6.3, col. 5). This value of r_e is also close to the median effective scale radii found in the Ostlin et al. (2001) study of nearby compact blue galaxies.

To compare the observed and predicted distributions, we use galactic models a–e from Bloom et al. (1999a) corresponding to hosts ranging in mass from $0.009 - 0.62 \times 10^{11} M_\odot$ and disk scale radii (r_e) of 1 and 3 kpc. Following the discussion above, we also construct a new model (a^*) which we consider the most representative of GRB hosts galaxies with $v_{circ} = 100$ km s^{-1}, $r_{break} = 1$ kpc, and $r_e = 1.5$ kpc ($M_{gal} = 9.2 \times 10^9 M_\odot$).

We project these predicted *radial* distribution models by dividing each offset by a factor of 1.15 since the projection of a merger site on to the plane of the sky results in a smaller observed distance to the host center than the radial distance. We determined the projection factor of 1.15 by a Monte Carlo simulation projecting a three-dimensional (3-D) distribution of offsets onto the sky. The median projected offset is thus 87% of the 3-D radial offset.

The observed distribution is compared with the predicted distributions and shown in figure 6.7 (later,

Table 6.4. Comparison of Observed Offset Distributions to Various Progenitor model Predictions

Progenitor	Comparison Model			Mass ($M_\odot$)	observed	P_{KS} synth	synth (replaced)	Fraction of $P_{KS} \geq 0.05$
	Name	r_e	$v_{\rm circ}$					
Collapsar… /Promptly Bursting Binary NS–NS, BH–NS…	expon. disk	$R_{\rm half}/1.67$			0.454	0.409	0.401	0.996
	a^*	1.5 kpc	100 km s^{-1}	9.2×10^9	9.5×10^{-4}	2.2×10^{-3}	2.2×10^{-3}	0.003
	a	1.0 kpc	100 km s^{-1}	9.2×10^9	6.9×10^{-3}	1.7×10^{-2}	1.8×10^{-2}	0.139
	b	1.0 kpc	100 km s^{-1}	2.8×10^{10}	4.9×10^{-3}	2.0×10^{-2}	2.0×10^{-2}	0.243
	c	3.0 kpc	100 km s^{-1}	2.8×10^{10}	3.5×10^{-5}	4.7×10^{-5}	4.7×10^{-5}	0.001
	d	1.0 kpc	150 km s^{-1}	6.3×10^{10}	2.5×10^{-2}	6.0×10^{-2}	6.3×10^{-2}	0.553
	e	3.0 kpc	150 km s^{-1}	6.3×10^{10}	5.7×10^{-5}	1.3×10^{-4}	1.3×10^{-4}	0.001

Note. — The names in column 2 correspond to population synthesis models from Bloom et al. (1999a). For the NS–NS, BH–NS comparisons, we believe that a^* best represents the observed host galaxy properties (see text). There are three columns that give the KS probability that the data are drawn from the model. The first, marked "observed," is the direct comparison of the models to the data without accounting for the uncertainties and measurements in the data. The second is the median KS probability derived from our Monte Carlo modeling (§6.C). The third is the median KS probability derived from our Monte Carlo modeling but where offsets thrown out due to high $P_{\rm chance}$ are replaced by synthetic offsets drawn from the model distribution. This pushes P_{KS} to larger values, in general, but the resulting median of the distribution is strongly affected (since $P_{\rm chance}$ is near zero for most offsets). The last column gives the fraction of synthetic datasets with $P_{KS} \geq 0.05$.

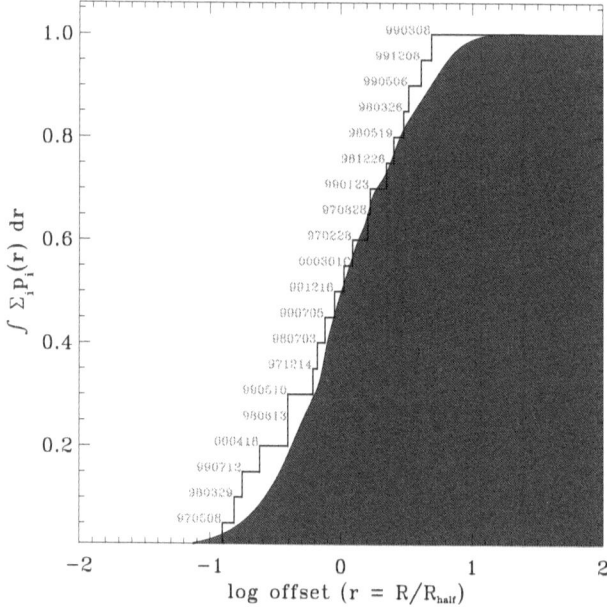

Figure 6.6 The cumulative GRB offset distribution as a function of host half-light radius. The solid jagged line is the data in histogram form. The smooth curve is the probability histogram (PH) constructed with the formalism of Appendix 6.B and is the integral of the curve depicted in figure 6.5. The GRB identifications are noted alongside the solid histogram. In this figure and in figure 6.5, SN 1998bw/GRB 980425 has not been included.

in fig. 6.8, we compare the observed distribution with the massive star prediction). We summarize the results in table 6.4. Only model d ($M = 6.3 \times 10^{10} M_\odot$, $r_e = 3$ kpc) could be consistent with the data ($P_{\rm KS} = 0.063$), but this galactic model has a larger disk and is probably more massive than most GRB hosts. Instead, for the "best bet" model a^*, the one-sided Kolmogorov-Smirnov probability that the observed sample derives from the same predicted distribution is $P_{\rm KS} = 2.2 \times 10^{-3}$, in agreement with our simplistic calculation above; that is, the location of GRBs appears to be inconsistent with the NS–NS and NS–BH hypothesis.

If GRBs do arise from systems which travel far from their birthsite, then there is a subtle bias in determining the offset to the host. If the progenitors are ejected from the host by more than half the distance between the host and the nearest (projected) galaxy, then the transient position will appear unrelated to any galaxies (the wrong host will be assigned, of course) but $P_{\rm chance}$ will always appear high no matter how deep the host search is. We try to account for this effect in our modeling (Appendix 6.C) by synthetically replacing observed (small) offsets that are associated with a high value of $P_{\rm chance}$ with new, generally larger, offsets drawing from the expected distribution of offsets for a particular galactic model. This then biases the distribution of P_{KS} statistics toward *higher* values (by definition), but the

6.7.2 Massive stars (collapsars) and promptly bursting binaries (BH–He)

As discussed earlier, collapsars produce GRBs in star-forming regions, as will BH–He binaries. The localization of GRB 990705 near a spiral arm is, of course, tantalizing smaller-scale evidence of the GRB–star-formation connection. Ideally, the burst sites of individual GRBs could be studied in detail with imaging and spectroscopy and should, if the collapsar/promptly bursting binary origin is correct, reveal that the burst sites are HII regions. Unfortunately, the distances to GRBs preclude a detailed examination of the specific burst sites on a resolution scale of tens of parsecs (the typical size for a star-forming region) with current instrumentation. Adaptive optics laser-guide star imaging may prove quite useful in this regard as will IR imaging with the *Next Generation Space Telescope*.

Weaker evidence for a star-formation connection exists in that no GRB to date has been observed to be associated with an early-type galaxy (morphologically or spectroscopically), though in practice it is often difficult to discern galaxy type with the data at hand. Indeed most well-resolved hosts appear to be compact star forming blue galaxies, spirals, or morphological irregulars.

Above we have demonstrated that GRBs follow the UV (restframe) light of their host galaxies. However, the comparison has been primarily mediated by a single parameter, the half-light radius and the median normalized offset. We now take this comparison one step further. For the GRB hosts with high signal-to-noise HST detections (e.g., GRB 970508, GRB 971214, GRB 980703), our analysis shows that the surface brightness is well-approximated by an exponential disk. We use this finding as the point of departure for a simplifying assumption about all GRB hosts: we assume an exponential disk profile such that the surface brightness of the host galaxy scales linearly with the galactocentric radius in the disk. We further assume that the star formation rate of massive stars scales with the observed optical light of the host; this is not an unreasonable assumption given that HST/STIS imaging probes restframe UV light, an excellent tracer of massive stars, at GRB redshifts.

Again, clearly not all host galaxies are disk-like (figure 6.2), so this assumption is not strictly valid in all cases. If r_e is the disk scale length, the half-light radius of a disk galaxy is $R_{\text{half}} = 1.67 \times r_e$, so that the simplistic model of the number density of massive star-formation regions in a galaxy is

$$N(r)\, dr \propto r \exp(-1.67\, r)\, dr, \tag{6.4}$$

where $r = R/R_{\text{half}}$. In reality, the distribution of massive star formation in even normal spirals is more complex, with a strong peak of star formation in the nuclear region and troughs between spiral arms (e.g., Rana & Wilkinson 1986; Buat et al. 1989). We make an important assumption when comparing the observed distribution with the star-formation disk model: that each GRB occurs in the disk of its host (see discussion below). Dividing the observed offset by the apparent half-light radius host essentially performs a crude de-projection.

We find the probability that the observed distribution could be derived from the simplistic distribution of massive star regions (equation 6.4) is $P_{\text{KS}} = 0.454$ (i.e., the two distributions are consistent). In Appendix 6.C we show that these results are robust even given the measurement uncertainties. This broad agreement between GRB positions and the UV light of their hosts is remarkable in the sense that the model for massive star locations is surely too simplistic; even in classic spiral galaxies (which most GRB hosts are not) star-formation is a complex function of galactocentric radius, with peaks in galactic centers and spiral arms. Furthermore, surface brightness dimming with redshift causes galaxies to appear more centrally peaked, resulting in a systematic underestimate of R_{half}

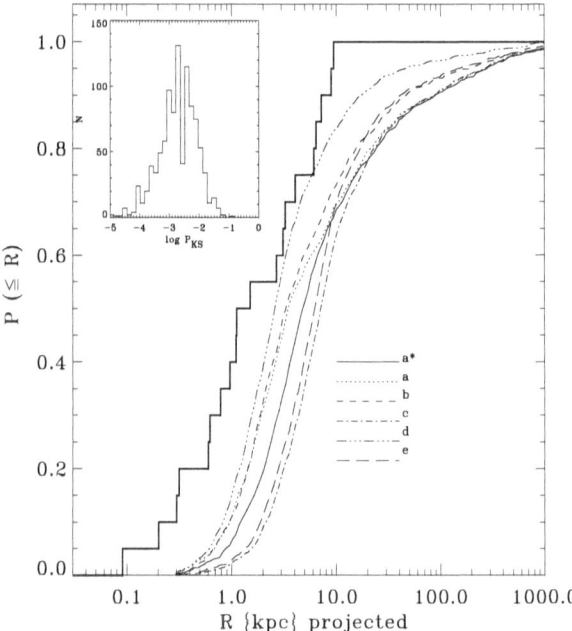

Figure 6.7 Offset distribution of GRBs compared with delayed merging remnant binaries (NS–NS and BH–NS) prediction. The models, depicted as smooth curves, are the radial distributions in various galactic systems that have been projected by a factor of 1.15 (see text). The letters denote the model distributions from table 2 of Bloom et al. (1999a); a^* is the galactic model which we consider as the most representative of GRB hosts galaxies ($v_{\rm circ} = 100$ km s^{-1}, $r_{\rm break} = 1$ kpc, $r_e = 1.5$ kpc, $M_{\rm gal} = 9.2 \times 10^9 M_\odot$). The cumulative histogram is the observed data set. Inset is the distribution of KS statistics (based on the maximum deviation from the predicted and observed distribution) of 1000 synthetic data sets compared with model a^*. Even with conservative assumptions (see text) the observed GRB distribution is inconsistent with the prediction: in only 0.3% of synthetic datasets is $P_{\rm KS} \geq 0.05$. Instead, the collapsar/promptly bursting remnant progenitor model appears to be a better representation of the data (see figure 6.8).

SECTION 6.8
Discussion and Summary

We have determined the observed offset distribution of GRBs by astrometrically comparing localizations of GRB afterglows with optical images of the field surrounding each GRB. In all cases, the GRB location appears "obviously" associated with a galaxy—either because the position is superimposed atop a galaxy or very near ($\lesssim 1.2''$) a galaxy in an otherwise sparse field. In fact, irrespective of the validity of individual assignments of hosts, the offset distribution may be considered a distribution of GRB positions from the nearest respective galaxy at least as bright as $R \approx 28$ mag (note that in most cases the host galaxies are much brighter, typically $R = 24\text{--}26$ mag). We find that at most a few of the 20 GRBs could be unrelated physically to their assigned host and about a 50% chance that all GRBs are correctly assigned to their hosts (see §6.6.1).

We then compare the distribution of GRB locations about their respective hosts with the *predicted* radial offset distribution of merging binary remnants. This comparison is complicated by an unknown projection factor for each burst: if a GRB occurs near an edge-on disk galaxy there exists no model-independent manner to determine the true 3-D radial offset of the GRB from the center of the host. Indeed, in a few cases (e.g., GRB 980519, GRB 991216) even the "center" of the host is not well defined and we must estimate a center visually. In all other cases, we find the centers using a luminosity-weighted centroid surrounding the central peak of the putative host.

To compare the GRB offsets with those predicted by the NS–NS and NS–BH binary models, we make a general assumption about the projection factor and, to facilitate a comparison in physical units (that is, offsets in kiloparsec rather than arcseconds), we assign an angular diameter distance to the 5 hosts without a confirmed distance (§6.6.2). We have shown that the conversion of an angular offset to physical projection is relatively insensitive to the actual redshift of the host. We estimate that the probability that the observed GRB offset distribution is the same as the predicted distribution of NS–NS and BH–NS binaries is $P \lesssim 2 \times 10^{-3}$. Insofar as the observed distribution is representative (see below) and the predicted distribution is accurate, our analysis renders BH–NS and NS–NS progenitor scenarios unlikely for long-duration GRBs.

Having cast doubt on the merging remnant hypothesis, we test whether the offset distribution is consistent with the collapsar (or BH–He) class. Since massive stars (and promptly merging binaries) explode where they are born, we have compared the observed GRB offset distribution with a very simplistic model of massive star formation in late-type galaxies: an exponential disk. After normalizing each GRB offset by their host half-light radius we compare the distribution with a KS test and find good agreement: $P_{\text{KS}} = 0.454$. We have shown that these KS results, based on the assumption of δ-function offsets, are robust even after including the uncertainties in the offset measurements.

Thus far we have neglected discussion of the observational biases that have gone into the localizations of these 20 GRBs. The usual problems plaguing supernova detection, such as the brightness of the central region of the host and dust obscuration, are not of issue for detection of the *prompt* high-energy emission (i.e., X-rays and γ-rays) of GRBs since the high-energy photons penetrate dust. If the intrinsic luminosity of GRBs is only a function of the inner-workings of the central engine (that is, GRBs arise from internal shocks and not external shocks) then the luminosity of a GRB is independent of ambient number density. Therefore, prompt X-ray localizations from BeppoSAX and γ-ray locations from the IPN should not be a function of the global properties of GRB environment; only intrinsic GRB properties such as duration and hardness will affect the prompt detection probability of GRBs.

The luminosity of the afterglows is, however, surmised to be a function of the ambient number density. Specifically, the afterglow luminosity will scale as $\sqrt{n}$ where n is the number density of hydrogen atoms in the 1–10 pc region surrounding the GRB explosion site (see Mészáros et al. 1998). While $n \approx 0.1\text{--}10$ cm^{-3} in the interstellar medium, the ambient number density is probably $n \approx 10^{-4}\text{--}10^{-6}$ in the intergalactic medium. Thus GRB afterglows in the IGM may appear $\sim 10^{-3}$ times fainter than GRB afterglows in the ISM (and even more faint compared to GRBs that occur in star-forming regions where the number densities are higher than in the ISM). If only a small fraction of GRBs localized promptly in X-rays and studied well at optical and radio wavelengths were found as afterglow, the ambient density bias may be cause for concern. However, this is not the case. As of June 2001, 29 of 34 bursts localized by prompt emission were later found as X-ray, optical, and/or radio afterglow (see Frail et al. 2000b); that is, almost all GRBs have detectable X-ray afterglow. Therefore, *no more than about 10% of GRBs localized by BeppoSAX could have occurred in significantly lower density environments* such as in the IGM; thus, we do not believe that our claim against the delayed merging binaries is affected by this bias.

What about the non-detection of GRB afterglow at optical/radio wavelengths? Roughly half of GRBs promptly localized in the gamma-ray or X-ray bands are not detected as optical or radio afterglow (Frail et al. 2000b). While many of these "dark" GRBs must be due to observing conditions (lunar phase, weather, time since burst, etc.) at least some fraction may be due to intrinsic extinction local to the GRB. If so, then these GRBs are likely to be centrally biased since the optical column densities are strongest in star-forming regions and giant molecular clouds. Therefore, *any optically obscured GRBs which do not make it in to our observed offset sample will be preferentially located in the disk*. We do not therefore believe the ambient density bias plays any significant role in causing GRBs to be localized preferentially closer to galaxies; in fact, the opposite may be true.

The good agreement between our simplistic model for the location of massive stars and the observed distribution is one of the strongest arguments yet for a collapsar (or promptly bursting binaries) origin of long-duration GRBs. However, the concordance of the predicted and observed distributions are necessary to prove the connection, although not sufficient.

We may now begin to relate the offsets to the individual host and GRB properties. For instance, of the GRBs which lie in close proximity to their host centers (GRB 970508, GRB 980703, and GRB 000418), there is a striking similarity between their hosts—all appear compact and blue with high-central surface brightness suggesting that these hosts are nuclear starburst galaxies (none show spectroscopic evidence for the presence of an AGN).

In fact, the closeness of some GRBs to their host centers signifies that our simplistic model for star-formation may require modification. This is not unexpected since, in the Galaxy, star formation as a function of Galactocentric radius does not follow a pure exponential disk, but is vigorous near the center and is strongly peaked around $R \sim 5$ kpc (see Kennicutt 1989). As more accurate offsets are amassed, these subtle distinctions in the GRBs offset distribution may be addressed.

The authors thank the staff of the W. M. Keck Foundation and the staff at the Palomar Observatories for assistance. We thank the anonymous referee for very insightful comments; in particular, the referee pointed out (and suggested the fix for) the bias in offset assignment if GRBs are ejected far from their host galaxies. The referee also pointed out that we did not establish that GRBs are preferentially aligned with the major axes of their hosts (as we had claimed in an earlier version of the paper). We applaud E. Berger and D. Reichart for close reads of various drafts of the paper. We also thank M. Davies for encouraging us to compare several NS-NS models to the data rather than just one. We acknowledge the members of the Caltech-NRAO-CARA GRB collaboration and P. van Dokkum, K. Adelberger, and R. Simcoe for helpful discussions. We thank N. Masetti for allowing us access to early ground-based data on GRB 990705 and B. Schaefer for kindly providing $QUEST$ images of the afterglow associated with GRB 990308. This work was greatly enhanced by the use of data taken as part of the *A Public Survey of the Host Galaxies of Gamma-Ray Bursts* with HST (#8640; S. Holland, P.I.). JSB gratefully acknowledges the fellowship and financial support from the Fannie and John Hertz Foundation. SGD acknowledges partial support from the Bressler Foundation. SRK acknowledges support from NASA and the NSF. The authors wish to extend special thanks to those of Hawaiian ancestry on whose sacred mountain we are privileged to be guests. Without their generous hospitality, many of the observations presented herein would not have been possible.

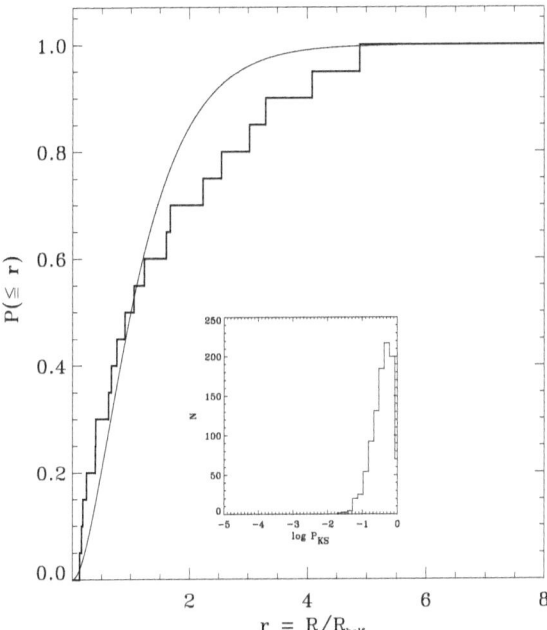

Figure 6.8 Offset distribution of GRBs compared with host galaxy star formation model. The model, an exponential disk, is shown as the smooth curve and was chosen as an approximation to the distribution of the location of collapsars and promptly bursting remnant binaries (BH–He). The cumulative histogram is the observed data set. Inset is the distribution of KS statistics (based on the maximum deviation from the predicted and observed distribution) of 1000 synthetic data sets. Since the observed KS statistic is near the median in both cases, we are assured that errors on the measurements do not bias the results of the KS test, and therefore the KS test is robust. The observed GRB distribution provides a good fit to the model considering we make few assumptions to perform the comparison. In reality the location of star formation in GRB hosts will be more complex than a simple exponential disk model.

SECTION 6.A
Potential Sources of Astrometric Error

6.A.1 Differential chromatic refraction

Ground-based imaging always suffers from differential chromatic refraction (DCR) introduced by the atmosphere. The magnitude of this refraction depends strongly ($\propto 1/\lambda_{\text{eff}}^2$) on the effective wavelength (λ_{eff}) of each object, the airmass of the observation, and the air temperature and pressure. With increasing airmass, images are dispersed by the atmosphere and systematically stretched in the parallactic direction in the sense that bluer objects shift toward the zenith and redder objects shift toward the horizon. Other sources of refraction, such as turbulent refraction (e.g., Lindegren 1980), are statistical in nature and will only serve to increase the uncertainty in our astrometric solution.

Here we show that DCR, in theory, will not dominate our offset determinations. Since all of our early ground-based imaging were conducted with airmass ($\sec(z)$) $\lesssim$ 1.6, we take as an extreme example an image with airmass $\sec(z) = 2$, where z is the observed zenith angle. It is instructive to determine the scale of systematic offset shifts introduced when compared with either late-time ground-based or HST imaging where refractive distortions are negligible. Following Gubler & Tytler (1998), the differential angular distortion between two point sources at an apparent angular separation along the zenith, Δz, may be broken into a color and a zenith distance term. Assuming nominal values for the altitude of the Keck Telescopes on Mauna Kea, atmospheric temperature, humidity and pressure, at an effective wavelength of the R-band filter, $\lambda_{\text{eff}}(R) = 6588$ Å (Fukugita et al. 1996), the zenith distance term is 16 mas for an angular separation of 30 arcsec at an airmass of $\sec(z) = 2$. The zenith term is approximately linear in angular distance and so, in practice, even this small effect will be accounted for as a first-order perturbation to the overall rotation, translation, and scale mapping between a Keck and HST image. In other words, we can safely neglect the zenith term contribution to the DCR.

We now determine the color term contribution. Optical transients of GRBs are, in general, redder in appearance (apparent $V - R \approx 0.5$ mag) than their host galaxies (apparent $V - R \sim 0.2$ mag). We assume the average astrometric tie object has $V - R = 0.4$ mag. If the OT is observed through an airmass of $\sec(z) = 2$ and then the galaxy is observed at a later time through and airmass of, for example, $\sec(z) = 1.2$, then DCR will induce a ~ 30 mas centroid shift between the OT and the host galaxy if the two epochs are observed in B-band (see figure 2 of Alcock et al. 1999). In R-band, the filter used in almost all of our ground-based imaging for the present work, the DCR strength is about 20% smaller than in B-band because of the strong dependence of refraction on wavelength. Therefore we can reasonably assume that DCR should only *systematically* affect our astrometric precision at the 5—10 mas level. Such an effect could, in principle, be detected as a systematic offset in the direction of the parallactic angles of the first epochs of GRB afterglow observations. In §6.6.1 we claim that no such systematic effect is present in our data. DCR could of course induce a larger *statistical* scatter in the uncertainty of an astrometric transformation between epochs since individual tie objects are not, in general, the same color and each will thus experience its own DCR centroid shift.

Bearing in mind that DCR is probably negligible we can minimize the effects of DCR by choosing small fields and similar spectral responses of the offset datasets. The HST fields are naturally small and there are enough tie stars when compared with deep ground-based imaging. However, since the spectral response of the HST/STIS CCD is so broad, extended objects with color gradients will have different apparent relative locations when compared with our deep ground-based R-band images. As such, in choosing astrometric tie objects, we pay particular attention to choosing objects which appear compact (half-light radii $\lesssim 0''.3$) on the STIS image.

6.A.2 Field distortion

Optical field distortion is another source of potential error in astrometric calibration. Without correcting for distortion in STIS, the maximum distortion displacement (on the field edges) is ~ 35 mas (Malumuth

& Bowers 1997). This distortion is corrected to a precision at the sub-milliarcsec level on individual STIS exposures with IRAF/DITHER (Malumuth & Bowers 1997). Malumuth & Bowers (1997) also found that the overall plate scale appears to be quite stable with r.m.s. changes at the 0.1% level. We confirmed this result by comparing two epochs of imaging on GRB 990510 and GRB 970508 which span about 1 year. The relative plate scale of the geometric mapping between final reductions was unity to within 0.03%.

We do not correct for optical field distortion before mapping ground-based images to HST. While there may be considerable distortion ($\sim$ few $\times 100$ mas) across whole ground-based CCD images, these distortions are correlated on small scales. Therefore, when mapping a 50×50 arcsec^2 portion of a Keck image with an HST image, the intrinsic differential distortions in the Keck image tend to be small ($\lesssim$ 30–50 mas). Much of the distortion is accounted for in the mapping by the higher-order terms of the fit, and any residual differential distortions simply add scatter to the mapping uncertainties.

SECTION 6.B
Derivation of the Probability Histogram (PH)

Histogram binning is most informative when there are many more data points than bins and the bin sizes are much larger than the errors on the individual measurements. Unfortunately, the set of GRB offsets is contrary to both these requirements. We require a method to display the data as in the traditional histogram, but where the errors on the measurements are accounted for. Instead of representing each measurement as a δ-function, we will represent each measurement as a probability distribution as a function of offset.

What distribution function is suitable for offsets? When the offset is much larger than the error, then the probability that the burst occurred at the measured displacement should approach a δ-function. When the offset is much larger than zero, then the probability distribution should appear essentially Gaussian (assuming the error on the measurement is Gaussian). However, when the observed offset is small and the error on the measurement non-negligible with respect to the observed offset, the probability distribution is decidedly non-Gaussian since the offset is a positive quantity. The distribution we seek is similar to the well-known Rice distribution (see Wax 1954), only more general.

We derive the probability histogram (PH) as follows. For each GRB offset, i, we construct an individual probability distribution function $p_i(r)\,dr$ of the host-normalized offset (r_i) of the GRB given the observed values for $X_{0,i}$, $Y_{0,i}$ and host half-light radius $R_{i,\text{half}}$ and the associated uncertainties. To simplify the notation in what follows, we drop the index i and let all lower case parameters represent dimensionless numbers; for example, the value $x_0 = X_0/R_{\text{half}}$, where R_{half} is the host half-light radius. Without loss of generality, we can subsume (by quadrature summation) the uncertainties in the host center, the astrometric transformation, and the GRB center into the error contribution in each coordinate. We assume that these statistical coordinate errors are Gaussian distributed with σ_x and σ_y with, for example,

$$\sigma_x = \frac{X_0}{R_{\text{half}}}\sqrt{\frac{\sigma_{X_0}^2}{X_0^2} + \frac{\sigma_{R_{\text{half}}}^2}{R_{\text{half}}^2}}.$$

Therefore, we can construct the probability $p(x,y)\,dx\,dy$ of the true offset at some distance x and y from the measured offset location (x_0, y_0):

$$p(x,y)\,dx\,dy = \frac{1}{2\pi\sigma_x\sigma_y}\exp\left[-\frac{1}{2}\left(\frac{x^2}{\sigma_x^2} + \frac{y^2}{\sigma_y^2}\right)\right]dx\,dy, \qquad (6.5)$$

assuming the errors in the x and y are uncorrelated. This is a good approximation since, while the astrometric mappings generally include cross-terms in X and Y, these terms are usually small. If $\sigma_x = \sigma_y$, then equation 6.5 reduces to the Rayleigh distribution in distance from the observed offset, rather than the host center.

The probability distribution about the host center is found with an appropriate substitution for x and y in equation 6.5. In figure 6.9 we illustrate the geometry of the problem. The greyscale distribution shows $p(x,y)\,dx\,dy$ about the offset point x_0 and y_0. Let $\phi = \tan^{-1}(y_0/x_0)$ and transform the coordinates in equation 6.5 using $\psi = \phi + \theta$, $x = r\cos\psi - x_0$, and $y = r\sin\psi - y_0$. The distribution we seek, the probability that the true offset lies a distance r from the host center, requires a marginalization of $\int_\psi p_i(r,\psi)\,dr\,d\psi$ over ψ,

$$\begin{aligned}p_i(r)\,dr &= \int_\psi p_i(r,\psi)\,dr\,d\psi \\ &= \frac{J\,dr}{2\pi\sigma_x\sigma_y}\int_0^{2\pi}\exp\left[-\frac{1}{2}\left(\frac{x(r,\psi)^2}{\sigma_x^2}+\frac{y(r,\psi)^2}{\sigma_y^2}\right)\right]d\psi,\end{aligned} \quad (6.6)$$

finding $J = r$ as the Jacobian of the coordinate transformation. In general, equation 6.6 must be integrated numerically using the observed values x_0, y_0, σ_x, and σ_y. The solution is analytic, however, if we assume that $\sigma_x \to \sigma_r$ and $\sigma_y \to \sigma_r$, so that

$$\begin{aligned}p_i(r)\,dr &\approx \frac{r}{\pi\sigma_r^2}\exp\left[-\frac{r^2+r_0^2}{2\sigma_r^2}\right]\int_{\theta=0}^{\pi}\exp\left[\frac{r\,r_0\cos\theta}{\sigma_r}\right]d\theta\,dr \\ &\approx \frac{r}{\sigma_r^2}\exp\left[-\frac{r^2+r_0^2}{2\sigma_r^2}\right]I_0\left(\frac{r\,r_0}{\sigma_r^2}\right)dr,\end{aligned} \quad (6.7)$$

where $I_0(x)$ is the modified Besel function of zeroth order and $r_0 = \sqrt{x_0^2+y_0^2}$.

The equation 6.7 is readily recognized as the Rice distribution and is often used to model the noise characteristics of visibility amplitudes in interferometry; visibility amplitudes, like offsets, are positive-definite quantities. Only when $\sigma_x = \sigma_y = \sigma_r$ is the probability distribution exactly a Rice distribution, which is usually the case for interferometric measurements since the real and imaginary components of the fringe phasor have the same r.m.s.

Equation 6.6 is a generalized form of the Rice distribution but can be approximated as a Rice distribution by finding a suitable value for σ_r. We find that by letting,

$$\sigma_r = \frac{1}{r_0}\sqrt{(x_0\,\sigma_x)^2+(y_0\,\sigma_y)^2}, \quad (6.8)$$

equation 6.7 approximates (to better than 30%) the exact form of the probability distribution in equation 6.6 as long as $\sigma_x \lesssim 2\,\sigma_y$ (or vise versa). In figure 6.10 we show two example offset probability distributions in exact and approximate form. Note that $r_0 - \sigma_r \leq r \leq r_0 + \sigma_r$ is not necessarily the 68% percent confidence region of the true offset since the probability distribution is not Gaussian. The exact form is used to construct the data representations in figures 6.6–6.8.

SECTION 6.C

Testing the Robustness of the KS Test

How robust are the estimates of probabilities found comparing the observed distribution and the predicted progenitor offset distributions? Since there are different uncertainties on each offset measurement, the KS test is not strictly the appropriate statistic to determine the likelihood that the observed distribution could be drawn from the same underlying (predicted) distribution. One possibility is to construct synthetic sets of observed data from the model using the observed uncertainties. However, a small uncertainty (say 0.2 arcsec in radius observed to be paired with an equally small offset) which is randomly assigned to a large offset from a Monte Carlo distribution has a different probability distribution then

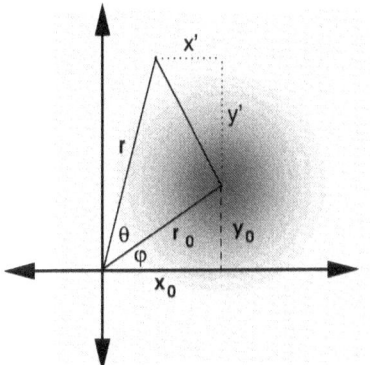

Figure 6.9 Geometry for the offset distribution probability calculation in Appendix 6.B.

if assigned to a small offset (since the distribution in r is only physical for positive r). Instead, we approach the problem from the other direction by using the data themselves to assess the range in KS statistics given our data. We construct $k = 1000$ synthetic cumulative physical offset distributions using the smoothed probability offset distributions $p_i(r)\,dr$ for each GRB. As before r is the offset in units of host half-light radius. For each simulated offset distribution k, we find a set $\{r_i\}_k$ such that

$$P[0,1] = \frac{\int_0^{r_i} p_i(l)\,dl}{\int_0^{\infty} p_i(l)\,dl},$$

where $P[0,1]$ is a uniform random deviate over the closed interval [0,1]. In addition, since some of the host assignments may be spurious chance superpositions, we use the estimate of P_chance (§6.6.1; table 6.3) to selectively remove individual offsets from a given Monte Carlo realization of the offset dataset. GRBs with relatively secure host assignments remain in more realizations than those without. So, for instance, the offset of GRB 980703 ($P_\text{chance} = 0.00045$) is used in all realizations but the offset of GRB 970828 ($P_\text{chance} = 0.07037$) is retained in only 93% of the synthetic datasets.

We evaluate the KS statistic as above for each synthetic set and record the result. Figure 6.8 depicts the cumulative probability distribution compared with the simple exponential disk model. The inset of the figure shows the distribution of KS statistics for the set of synthetic cumulative distributions constructed as prescribed above. In both cases, as expected, the *observed* KS probability falls near the median of the synthetic distribution. The distribution of KS statistics is not significantly affected by retaining all GRB offsets equally (that is, assuming $P_\text{chance} = 0.0$ for every GRB offset). In table 6.4 we present the result of the Monte Carlo modeling. Using this distribution of KS statistics we can now assess the robustness of our comparison result: given the data and their uncertainties, the probability that the observed GRB offset distribution is the same as the model distribution of star formation (exponential disk) is $P_\text{KS} \geq 0.05$ in 99.6% of our synthetic datasets.

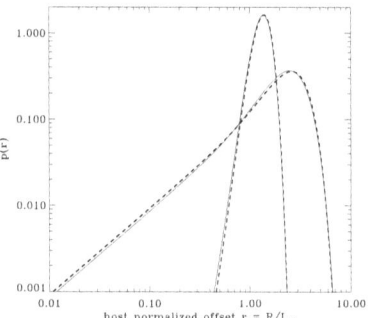

Figure 6.10 Example offset distribution functions $p(r)$. Depicted are two probability distribution curves for $(X_0, Y_0, \sigma_{X_0}, \sigma_{Y_0}, R_{\text{half}}, \sigma_{R_{\text{half}}}) = [0''.033, 0''.424, 0''.034, 0.''034, 0''.31, 0''.05]$ (GRB 970228) and $[0''.616, 0''.426, 0''.361, 0''.246, 0''.314, 0''.094]$ (GRB 981226) for the lower and upper peaked distributions, respectively. The solid line is the exact solution (equation 6.6) and the dashed line is the approximate solution (equation 6.7). Here, as in the text, the host-normalized offset $r = R/R_{\text{half}}$, where R is the galactocentric offset of the GRB from the host and R_{half} is the half-light radius of the host.

Part II

The GRB/Supernova Connection

I never cared much for moonlit skies
I never wink back at fireflies
But now that the stars are in your eyes
I'm beginning to see the light

I never went in for afterglow
Or candlelight on the mistletoe
But now when you turn the lamp down low
I'm beginning to see the light

"I'm Beginning To See The Light," 1944
written by
Johnny Hodges, Don George, Harry James, & Duke Ellington
performed by Ella Fitzgerald, Duke Ellington Songbook, disc 5

CHAPTER 7

The Unusual Afterglow of the Gamma-ray Burst of 26 March 1998 as Evidence for a Supernova Connection[†]

J. S. BLOOM[1], S. R. KULKARNI[1], S. G. DJORGOVSKI[1], A. C. EICHELBERGER[1], P. CÔTÉ[1], J. P. BLAKESLEE[1], S. C. ODEWAHN[1], F. A. HARRISON[1], D. A. FRAIL[2], A. V. FILIPPENKO[3], D. C. LEONARD[3], A. G. RIESS[3], H. SPINRAD[3], D. STERN[3], A. BUNKER[3], A. DEY[4], B. GROSSAN[6], S. PERLMUTTER[7], R. A. KNOP[7], I. M. HOOK[8], & M. FEROCI[9]

[1]Palomar Observatory 105-24, Caltech, Pasadena, CA 91125, USA

[2]National Radio Astronomy Observatory, P. O. Box O, Socorro, NM 87801, USA

[3]Department of Astronomy, University of California, Berkeley, CA 94720-3411 USA

[4]National Optical Astronomy Observatories, 950 N. Cherry, Ave. Tucson, AZ 85719, USA

[6]Center for Particle Astrophysics, University of California, Berkeley, CA 94720 USA

[7]Lawrence Berkeley National Laboratory, Berkeley, CA 94720, USA

[8]European Southern Observatory, D-85748 Garching, Germany

[9]Istituto di Astrofisica Spaziale, CNR, via Fosso del Cavaliere, Roma I-00133, Italy

Abstract

Cosmic gamma-ray bursts have now been firmly established as one of the most powerful phenomena in the universe, releasing almost the rest-mass energy of a neutron star in a few seconds (Kulkarni et al. 1999a). The most popular models to explain gamma-ray bursts are the coalescence of two compact objects such as neutron stars or black holes, or the catastrophic collapse of a massive star in a very energetic supernova-like explosion (MacFadyen & Woosley 1999; Paczyński 1998). An unavoidable consequence of the latter model is that a bright supernova should accompany the GRB. The emission from this supernova competes with the much brighter afterglow produced by the relativistic shock that gives rise to the GRB itself. Here we show that about 3 weeks after the gamma-ray burst of 26 March 1998, the transient optical source associated with the burst brightened to about 60 times the expected flux, based upon an extrapolation of the initial light curve. Moreover, the spectrum changed dramatically, with the color becoming extremely red. We argue that the new source is an underlying supernova. If our hypothesis is true then this provides evidence linking cosmologically located gamma-ray bursts with deaths of massive stars.

[†] A version of this chapter was first published in *Nature*, 401, p. 453–456, (1999).

Section 7.1
Introduction

The origin of GRBs remained elusive for a period of nearly three decades after their discovery (Klebesadel et al. 1973). Beginning in 1997, however, the prompt localization of GRBs by the Italian-Dutch satellite BeppoSAX (Boella et al. 1997) and the All Sky Monitor (Levine et al. 1996) on board the X-ray Timing Explorer led to the discovery of the GRB afterglow phenomenon – emission at lower energies: X-ray (Costa et al. 1997a), optical (van Paradijs et al. 1997), and radio (Frail et al. 1997).

The persistence of the afterglow emission (days at X-ray wavelengths, weeks to months at optical wavelengths, months to a year at radio wavelengths) enabled astronomers to carry out detailed observations which led to fundamental advances in our understanding of these sources: (1) the demonstration that GRBs are at cosmological distances (Metzger et al. 1997b); (2) the proof that these sources expand with relativistic speeds (Frail et al. 1997); and (3) the realization that the electromagnetic energy released in these objects exceeds that in supernovae (Waxman et al. 1998) and, in some cases, the released energy is comparable to the rest mass energy of a neutron star (Kulkarni et al. 1998b; Djorgovski et al. 1998; Kulkarni et al. 1999a; Anderson et al. 1999).

Despite these advances, we are still largely in the dark about the nature of the GRB progenitors. Though there are a number of models for their origin, the currently popular models involve the formation of black holes resulting from either the coalescence of neutron stars (Paczyński 1986; Goodman 1986; Narayan et al. 1992) or the death of massive stars (Woosley 1993; Paczyński 1998). The small offsets of GRBs with respect to their host galaxies and the association of GRBs with dusty regions and star-formation regions favors the latter, the so-called hypernova scenario (Paczyński 1998). However, this evidence is indirect and also limited by the small number of well-studied GRBs (see, however, chapter 6 which was published after this chapter).

The most direct evidence for a massive star origin would be the observation of a supernova coincident with a GRB. Here we present observations of GRB 980326 and argue for the presence of such an underlying supernova. If our conclusions are correct, then the implication is that at least some fraction of GRBs, perhaps the entire class of long duration GRBs, represent the endpoint of the most massive stars. Furthermore, if the association (Galama et al. 1998b; Kulkarni et al. 1998) of GRB 980425 with a bright supernova in a nearby galaxy holds, then the apparent γ-ray luminosity of GRBs ranges over six orders of magnitude.

Section 7.2
The Unusual Optical Afterglow

Following the localization of GRB 980326 by BeppoSAX (Celidonio et al. 1998), Groot et al. (1998a) quickly identified the optical afterglow. Our optical follow-up program began at the Keck Observatory, approximately 10 hr after the burst. A log of these observations is given in table 7.1.

In figure 7.2 we present our R-band photometry, along with values reported by other workers. Considering only reported data taken within the first month of the burst, we find a characteristic power law decay in the flux versus time, followed by an apparent flattening. The usual interpretation is that the decaying flux is the afterglow emission, while the constant flux is due to the host galaxy. Indeed, earlier (Djorgovski et al. 1998) we attributed the entire observed flux on April 17th to the host galaxy.

But to our surprise, our more recent observations (first performed nine months after the GRB event) showed no galaxy at the position of the optical transient (OT); see figure 7.1. We estimate a 2-σ upper limit the R-band magnitude of $R > 27.3$ mag (see table 7.1). This is almost a factor of 10 less flux than that reported from our 17 April detection. A secure conclusion is that the presumed host galaxy of GRB 980326, assuming that the GRB was coincident with the host (as appears to be the case for all other well-studied GRBs to date), is fainter than $R \approx 27$ magnitude. This conclusion is not alarming as such faint (or fainter) galaxies are indeed expected from studies (Mao & Mo 1998; Hogg & Fruchter

Table 7.1. Keck II Optical Observations of GRB 980326

Date[a] (UT)	Band/ Grating	Int. Time (sec)	Seeing (FWHM)	Magnitude	Observers
Mar 27.35	R	240	$0''.74$	21.25 ± 0.03	AVF, DCL, AGR
Mar 28.25	R	240	$0''.66$	23.58 ± 0.07	HS, AD, DS, SAS
Mar 29.27	300	3600		24.45 ± 0.3	HS, AD, DS, SAS
Mar 30.24	R	900	$0''.93$	24.80 ± 0.15	SP, BG, RK, IH
Apr 17.25	R	900	$0''.82$	25.34 ± 0.33	PC, JB
Apr 23.83	300	5400		24.9 ± 0.3^c	SGD, SCO
Dec 18.50	R	2400	$0''.74$	> 27.3	SRK, JSB, MvK
Dec 18.54	I	2100	$0''.74$	> 25.3	SRK, JSB, MvK
Mar 24	I	5450	$0''.80$	> 26.6	SRK, JSB

Note. — We used the Keck II 10-m Telescope 2,048 × 2,048 pixel CCD (charged coupled device) Low-Resolution Imaging Spectrometer (LRIS; Oke et al. 1995) for imaging and spectroscopy of the GRB field. The epoch of GRB 980326 is 26.888 March 1998 (Celidonio et al. 1998). For details of the data reductions, see §7.A.

[a]Mean epoch of the image. The year is 1998 for all images except for that on March 24 for which it is 1999.

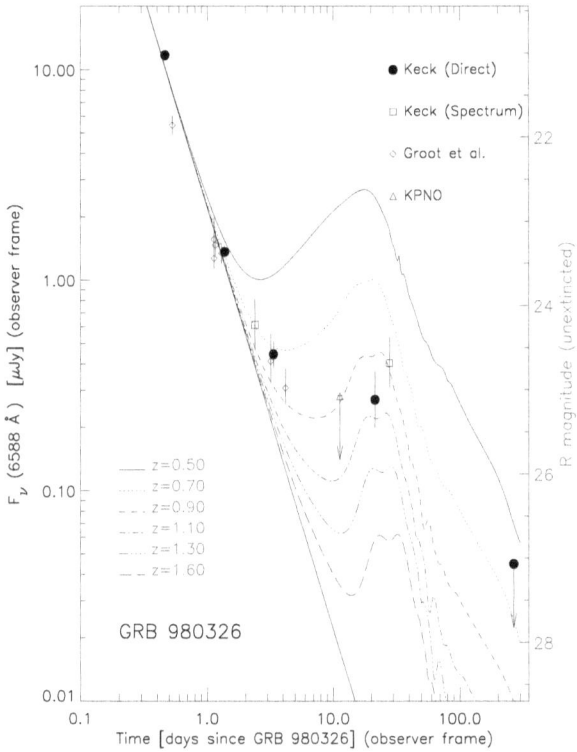

Figure 7.1 The R-band light curve of the afterglow of GRB 980326. Overlaid is a power-law afterglow decline summed with a bright supernova light curve at different redshifts. (Although we use as as a template the multi-band light curve of SN 1998bw (Galama et al. 1998b; McKenzie & Schaefer 1999), the bright supernova potentially associated with GRB 980425, we emphasize that the exact light-curve shapes of a supernova accompanying a GRB is not known *a priori*.) The GRB+SN model at redshift of about unity provides an adequate description of the data. See §7.B for future details.

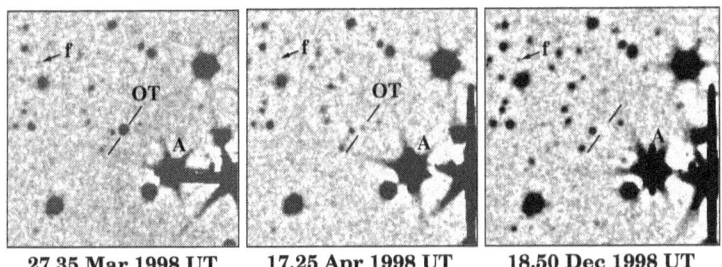

27.35 Mar 1998 UT   17.25 Apr 1998 UT   18.50 Dec 1998 UT

Figure 7.2 Images of the field of GRB 980326 at three epochs. Each images shows a $54'' \times 54''$ region centered on the optical transient (labeled "OT"). In all the images, the local background has been subtracted by a median filter and the resulting image smoothed (with a two-dimensional Gaussian with $\sigma = 0''.23$). An unrelated faint source "f" in the field is noted for comparison of the relative limiting flux between the three epochs: it is marginally detected (at the $\sim 2\text{-}\sigma$ level) on March 27 and April 17 but well detected on December 18. In contrast the OT is brighter and better detected (at the 4.6-σ level, see text and §7.C) on April 17 but clearly not detected to fainter levels on December 18 ($R > 27.3$ mag; see table 7.1).

1999) of the properties of cosmological GRB host galaxies.

Having established that the host galaxy of GRB 980326 is faint, we are forced to conclude that the OT did not continue the rapid decay it exhibited initially. Instead, we find three phases of the light curve (figure 7.2): a steeply declining initial phase (at times since the burst of $t \lesssim 5$ day), a subsequent re-brightening phase ($t \sim 3$–4 weeks) and, finally, a phase in which the source appears to have faded away to an undetectable level by the time of our next observation (9 months after the burst).

In previously studied bursts, the optical afterglow emission has been modeled by a power-law function, flux $\propto t^\alpha$; here α is the power-law index. In some bursts, at early times (t less than a day or so), significant deviations have been seen, for example, GRB 970508 (Djorgovski et al. 1997b). At late times, in some bursts, deviations manifest as steepening (that is, α becoming smaller) of light curves, for example, GRB 990123 (Kulkarni et al. 1999a) and GRB 990510 (Harrison et al. 1999; Stanek et al. 1999).

It is against this backdrop of the observed afterglow phenomenology that we now analyze the light curve in figure 7.2. The declining phase cannot be fitted by a simple power law ($\chi^2 = 72$ for 9 degrees of freedom). From figure 7.2 it is clear that the flux had already started flattening by day 3. Restricting the analysis to the first two days, we obtain $\alpha = -2.0 \pm 0.1$, consistent with a previous analysis (Groot et al. 1998a).

Such power-law decays are usually interpreted as arising from electrons shocked by the explosive debris sweeping up the ambient medium (Katz 1994; Mészáros & Rees 1997a; Vietri 1997; Waxman 1997b). Assuming that the electrons behind the shock are accelerated to a power-law differential energy distribution with index $-p$, on general grounds (Sari et al. 1998) we expect that the afterglow flux, $f_\nu(t)$, is proportional to $t^\alpha \nu^\beta$; here $f_\nu(t)$ is the flux at frequency ν and time t. The value of α and β depend on p, the geometry of the emitting surface (Mészáros et al. 1998) (spherical versus collimation) and the radial distribution of the medium around the burst (Chevalier & Li 1999).

From our spectroscopic observations of 29 March (figure 7.3), we find $\beta = -0.8 \pm 0.4$. This combination of (α, β) is similar to the $(\alpha = -2.05 \pm 0.04, \beta = -1.20 \pm 0.25)$ seen in GRB 980519 (Halpern et al. 1999), and can be reasonably interpreted (Sari et al. 1999) as arising from a standard $p \sim 2.2$ shock with a jet-like emitting surface. Alternatively, the emission could arise in a $p \sim 3$ shock propagating in

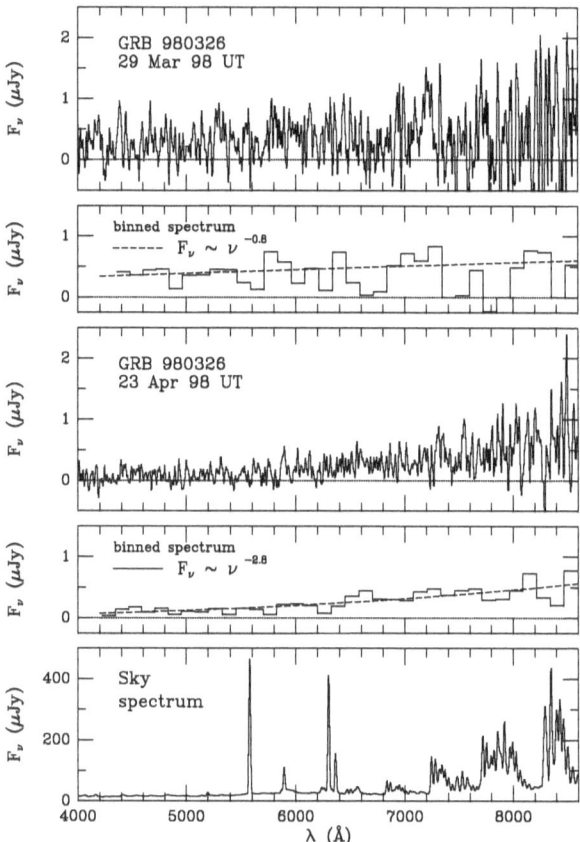

Figure 7.3 The spectra of the transient on March 29.27 and April 23.83 1998 UT. The two spectra are shown at two different spectral resolutions. Starting from the top, panels 1 and 3 show the spectra at the two epochs at the full spectral resolution (see below for details) and panels 2 and 4 show the same two spectra but binned in groups of 51 channels. Panel 5 is the spectrum of the sky. The best-fit power law models ($f_\nu \propto \nu^\beta$) to the binned spectra are shown by dashed lines; the fits were restricted to the wavelength range 4500–8500 Å. The scatter of individual channel values within each bin was used to assign relative weights to the median fluxes in each bin when performing the fits. On March 29.27, we obtain $\beta = -0.8 \pm 0.4$ and $\beta = -2.8 \pm 0.3$ on April 23.83. The derived power-law indices include the correction of Galactic extinction. From the absence of continuum breaks in the spectrum of March 29, we can place an upper limit to the redshift, $z_{OT} \lesssim 2.3$. See §7.D for details.

a circumburst medium (Chevalier & Li 1999) whose density falls as the inverse square of the distance from the explosion site.

SECTION 7.3
A New Transient Source

We now discuss the bright source seen in the re-brightening phase (corresponding to observations of April 17 and April 23). This source is ~ 60 times brighter than that extrapolated from the rapidly declining afterglow. Given that a magnitude of excess at late times has not been reported before, it is important to review the crucial observation of mid-April.

First, the re-brightened source is coincident with the OT in the image of 27 March to within the expected astrometric error (0.04 ± 0.18 arcsecond). As noted in the legend to figure 7.1, the source is consistently detected in three separate frames taken on 17 April. In the summed image, the source is detected at 4.6-σ (chance probability of 2×10^{-6}) on 17 April and all other objects in the field at this flux level are reliably detected in our deeper 18 December image. Next, the source is clearly detected (spectroscopically) on 23 April (figure 7.3) at the same position as that of the OT. The inferred spectrophotometric R-band magnitude is plotted as open squares on the light curve (figure 7.2). Thus we conclude that there was indeed a source at the position of the OT which brightened three weeks after the burst and subsequently faded to undetectable levels. We now investigate possible explanations for this source.

The simplest picture is that afterglow re-brightened. Piro et al. (1999) have recently suggested that the doubling of the X-ray flux of GRB 970508 three days after the GRB event arises from the relativistic shell running into a dense gas cloud. Such an explanation for the GRB 980326 light curve would require a large dense region, with a size comparable to the timescale of re-brightening, about $\Gamma \times 10$ light days (~ 0.01 pc) and located at a distance $\Gamma^2 c \times 20$ days (~ 0.1 pc) from the explosion site. Here, Γ is the bulk Lorentz factor of the shock and is expected to be order unity three weeks after the burst. Panaitescu et al. (1998) suggest that the re-brightening of GRB 970508 may be due to a shock refreshment– delayed energy injection by the extremely long-lived central engine that produced the GRB. Alternatively, the delayed energy could come from the spin-down of a newly formed milli-second pulsar through magnetic dipole radiation Dai & Lu (2000). For all these models, however, the expected spectrum would be the typical synchrotron spectrum, flux $f_\nu \propto \nu^{-1}$ (or flatter). The very red spectrum of 21 April (figure 7.3) allows us to essentially rule out a synchrotron origin for the re-brightening phase of GRB 980326.

Alternatively, as suggested by Loeb (1999), the GRB could have occurred in a dusty region and the afterglow would re-brighten as the dust is sublimated by the afterglow. However, the observed spectral evolution from a relatively blue spectrum (29 March) to red (23 April) moves in a direction opposite to that expected in this model. Last, a non-relativistic, thermal expanding envelope powered by radioactive decay could accompany the merger of a compact binary system (Li & Paczyński 1998) that gives rise to the GRB and the afterglow: this envelop would radiate at luminosities comparable to that of a bright supernova and should produce a red spectrum similar to that seen on 23 April; however, the time-scale for the peak emission is more than an order of magnitude shorter than the timescale for the re-brightening that we see with GRB 980326.

SECTION 7.4
The Supernova Interpretation

We advance the hypothesis that the new source is due to an underlying supernova revealed only after the afterglow emission has vanished. Woosley and collaborators (MacFadyen & Woosley 1999; Woosley 1993 and references therein) have pioneered the "collapsar" model in which GRBs arise from the death of massive stars—stars which produce black hole remnants rather than neutron stars. In this model, the iron core of a massive star collapses to a black hole and releases up to a few $\times 10^{52}$ erg of kinetic energy. Some fraction of this energy is expected to emerge in the form of a jet with little entrained matter; bursts of gamma-rays result from internal shocks in this jet. The remaining energy is absorbed by the star, causing it to explode and thereby produce a supernova (Hansen 1999).

Thus in this model, the total light curve has two distinct contributions: a power-law decaying afterglow component, and emission from the underlying supernova. In figure 7.2 we show the light curve expected in this model and use the light curve of the well observed (Galama et al. 1998b; McKenzie & Schaefer 1999) SN 1998bw as a template for the supernova contribution. We find the R-band and I-band data to be consistent with a bright supernova at $z \approx 1$.

The very red spectrum of the source on 23 April finds a natural explanation in the supernova hypothesis. On theoretical (MacFadyen & Woosley 1999) and phenomenological (Kulkarni et al. 1998) grounds, we expect GRBs to arise from massive stars which have lost their hydrogen envelope, that is, type Ibc supernovae. At low redshifts, all type I supernovae are observed to exhibit a strong ultraviolet deficit relative to the blackbody fit to their spectra. This deficit is due to absorption by prominent atomic resonance lines starting below ~ 3900 Å. Below $\lambda_c \sim 2900$ Å we expect to see very little flux. In the near-ultraviolet range (3000–4000 Å) all spectra from type I supernovae have a red appearance. Approximating the flux by a power law ($f_\nu \propto \nu^\beta$), the ultraviolet power-law index (depending on the wavelength range chosen) is -3 or even smaller, as found in the prototypical type Ic SN 1994I (Kirshner et al. 1993). Fitting the spectrum of figure 7.3 to a power law, we obtain $\beta = -2.8 \pm 0.3$. Such a red spectrum (negative β) requires that the observed spectrum corresponds to the ultraviolet–blue region in the restframe of the object. A smaller redshift would lead to a larger β. A larger redshift would substantially suppress the light in the observed R band (which covers the wavelength range 5800–7380 Å). The detection in R band then provides an independent constraint (figure 7.2), $z \lesssim 1.6$.

We have used the light curve of SN 1998bw because it is a very well studied type Ibc SN with a possible association with GRB 980425. We do not know *a priori* the precise spectrum and light curve of a supernova accompanying GRBs. In the collapsar model, the progenitor stars are expected to have no significant envelopes and thus the expected supernovae are of type Ibc. The general shape of the spectra of all type Ibc supernovae are expected to be the same and are summarized above. Given the low signal-to-noise ratio of the 23 April spectrum and the expected line broadening due to high photospheric velocity, we do not, as seems to be the case, expect to see any features in our spectrum.

Independently, from the absence of strong spectral breaks in our spectrum of the OT, we can firmly place the redshift of the OT at $\lesssim 2.3$. This constraint is consistent with our deduction that $z \lesssim 1.6$ (see above). Thus from a variety of accounts we find a plausible redshift of around unity for GRB 980326. Such a redshift is not entirely unexpected. Indeed, we note that five out of eight spectroscopically confirmed redshifts of GRBs lie in the range $0.7 < z < 1.1$.

SECTION 7.5
Implications of the Supernova Connection

The direct evidence for an accompanying SN can be seen in the light curve at timescales comparable to the time for SNe to peak, $\sim 20(1+z)$ days. However, in our opinion three conditions must be satisfied in order to see the underlying SN even when one was present. (1) The GRB afterglow should decline rapidly, otherwise the SN will remain overpowered by the afterglow for all epochs. (2) Given the strong ultraviolet absorption (discussed above), only GRBs with redshift $z \lesssim 1.6$ have an observable SN component in the optical band. (3) The host must be dimmer than the peak magnitude of the SN ($M_V \sim -19.5$ mag). The last requirement is not needed if the GRB can be resolved from the host (for example, by using the Hubble Space Telescope). Finally, one caveat is worth noting: the peak magnitudes of Type Ibc SNe are not constant (unlike those of Type Ia), and can vary (Iwamoto et al. 1998) from -16 mag to a maximum of -19.5 mag (see Germany et al. 2000). We have investigated the small sample of GRBs with adequate long-term follow up and conclude that perhaps only GRB

980519 satisfies the first and the third observational conditions for supernova detection; the redshift of this GRB is unfortunately unknown.

The dynamics of the relativistic blast wave is strongly affected by the distribution of circumstellar matter. Chevalier & Li (1999) note that massive stars, through their active winds, leave a circumstellar medium with density falling as the inverse square of the distance from the star. One expects smaller α for GRBs exploding such a circumstellar medium. In this framework, GRB afterglows which decline rapidly and are at modest redshifts will again be prime targets to search for the underlying SN.

If we accept the SN interpretation for GRB 980326, a long-duration (5 sec) GRB, then it is only reasonable to posit that all other long duration GRBs are also associated with supernovae. We suggest that sensitive observations be made—especially at longer wavelengths, to avoid the UV cutoff of supernovae—of GRBs satisfying the above three conditions. If our proposed hypothesis is correct, then the light curves and the spectra of such GRBs would exhibit the behavior shown in figure 7.2 and figure 7.3 and discussed here. Indeed, motivated by this work, evidence for underlying supernovae in other GRBs is now being reported (GRB 970228; Reichart 1999).

We end with a discussion of one interesting point. The total energy release in γ-rays of GRB 980326 was $E_\gamma = (3.42 \pm 3.74) \times 10^{51} f_b$ erg where f_b is the fractional solid angle of the jet (if any); here we have used the measured fluence (Groot et al. 1998a) and assumed $z \sim 1$ ($H_0 = 65$ km s^{-1} Mpc^{-2}, $\Omega_0 = 0.3$, $\Lambda_0 = 0.7$) (see also Bloom et al. 2001b). If this GRB was beamed, then $E_\gamma \sim 10^{49}$ erg. Curiously enough, this rather small energy requirement places GRB 980326 as close in energetics to GRB 980425 ($E_\gamma = 7.16 \times 10^{47}$ erg; Galama et al. 1998b and chapter 9) as to the classic gamma-ray bursts ($E_\gamma \gtrsim 5 \times 10^{50}$ erg; Frail et al. 2001).

Note added in proof: Galama et al. (2000) have also recently reported supernova-like behavior in the light curve and spectrum of 970228.

Acknowledgment. We thank M. H. van Kerkwijk for help with the December 18 observations at the Keck II telescope and R. Sari for helpful discussions. We gratefully acknowledge the excellent support from the staff at the Keck Observatory. The observations reported here were obtained at the W. M. Keck Observatory, made possible by the generous financial support of the W. M. Keck Foundation, which is operated by the California Association for Research in Astronomy, a scientific partnership among California Institute of Technology, the University of California and the National Aeronautics and Space Administration. SRK's and AVF's research is supported by the National Science Foundation and NASA. SGD acknowledges partial support from the Bressler Foundation.

SECTION 7.A

Details of the Data Reduction for Table 7.1

7.A.1 Photometric calibration

The absolute zero-point of the R (effective wavelength of $\lambda_{\text{eff}} \approx 6588$ Å; Fukugita et al. 1995) and I-bands ($\lambda_{\text{eff}} \approx 8060$ Å) were calibrated to the standard Cousins bandpass using standard-stars in the field SA98 (Landolt 1992) and assuming the standard atmospheric correction on Mauna Kea (0.1 mag and 0.06 mag per unit airmass, respectively). The estimated statistical error on the absolute zero-point is 0.01 mag. We estimate the systematic error (due to lack of inclusion of color term) to be less than 0.1 mag. We propagated all photometry to the absolute zero-point derived in the first epoch of observation using 8 "secondary" stars which were detected with high signal-to-noise ratio, unsaturated, near to the transient, and common to every epoch; the typical uncertainty in the zero-point propagation is 0.01 mag. Thus any systematic error in our absolute zero-point will not affect the conclusions based on *relative* flux. The uncertainties quoted in the table 7.1 contain all known sources of error (aperture correction, etc.). The calibrated magnitudes of the secondary stars reported in Groot et al. (1998a) agree to within the measurement errors.

7.A.2 Spectrophotometric measurement

The flux in µJy is determined at 6588 Å, the central wavelength of the R_c band; the conversion to magnitude assumes 0 mag equal to 3020 Jy (Fukugita et al. 1995). The spectrophotometric magnitudes are relative to a bright star that was on the slit (for which we have obtained independent photometry from our images).

7.A.3 Photometry of the faint source

Since the transient was not detected to significantly fainter levels in later epochs, it is safe to assume that the April 17 detection was that of a point-source (and not an extended galaxy as we had earlier believed; Djorgovski et al. 1998). To maximize the signal-to-noise ratio, we choose to measure the photometry in an aperture radius equal to the FWHM of the seeing and correct for the missing flux outside the aperture by using the radial flux profiles of bright isolated stars in the image. The determination of the optimum sky level (from which we subtract the total flux in the aperture) is not well-defined. We estimate the systematic uncertainty introduced by the uncertainty in the sky level as 0.25 mag. The statistical uncertainty (weighted mean over different background determinations) of the flux was 0.22 mag. Thus we quote the quadrature sum of the statistical and systematic uncertainty of 0.33 mag.

7.A.4 Upper-limits

On 1998 Dec 18 UT and 1999 March 24 UT there was no detectable flux above the background at the position of the optical transient. We centered 1000 apertures randomly in our image (approximately 1800×2048 pixels in size) and performed weighted aperture photometry with a local determination of sky background and recorded the counts ("DN") above background at each location. The flux contribution from an individual pixel, some radius r from the center of the aperture, to the total flux was weighted by a Gaussian with a radial width FWHM equal to the seeing. A histogram of the resulting flux was constructed. This histogram was decomposed into two components—a Gaussian with median near zero DN and a long tail of positive DN corresponding to actual source detections. We fit a Gaussian to the zero-median component, iteratively rejecting outlier aperture fluxes. Based on the photometric zero-point and using isolated point sources in the image for aperture corrections, we computed the relationship between DN within the weighted aperture and the total magnitude. In table 7.1 we quote an upper limit (95%-confidence level corresponding to 2-σ of the Gaussian fit) at the position of the optical transient.

SECTION 7.B
Notes on Figure 7.2

7.B.1 Transient light curve

From Schlegel et al. (1998) we estimate the Galactic extinction in the direction of the optical transient $(l, b = 242°.36, 13°.04)$ to be $E(B - V) = 0.08$. Thus, assuming the average Galactic extinction curve $(R_V = 3.1)$, the extinction measure is $A_R = 0.22$ mag, $A_I = 0.16$ mag. Plotted are the extinction corrected magnitudes (see table 7.1) of the transient converted to the standard flux zero-point of the Cousins R filter from Fukugita et al. (1995). In addition to our data, we include photometric detections from Groot et al. (1998a) and an upper-limit from Valdes et al. (1998) (KPNO). The GRB transient flux dominates at early times, but with a power-law decline slope $\alpha = -2$ (straight solid line).

7.B.2 Supernova light curve

The supernova light curve template was constructed by spline-fitting the broadband spectrum measured by Galama et al. (1998b) of the bright supernova 1998bw at various epochs (augmented with late-time observations of SN 1998bw by McKenzie & Schaefer 1999) and transforming back to the restframe of SN 1998bw ($z = 0.0088$). As discussed in the text, we expect the rest-frame UV emission (below 3900Å) to be suppressed due to absorption by resonance lines. We assume that the UV flux declines as $f_\nu \propto \nu^{-3}$. Theoretical light curves are then constructed by red-shifting the template to various redshifts and determining the flux in the R (observer frame) by interpolating (or extrapolating, for $z \gtrsim 1$). The flux normalization of the redshifted SN 1998bw curves are independent of the Hubble constant but are dependent upon the value of Ω_0 and Λ_0 (here we show the curves for $\Omega_0 = 0.2$ and $\Lambda_0 = 0$). Beyond $z \approx 1.3$ the observed R-band corresponds to restframe $\lambda \lesssim 2900$ Å. As stated in the text, the spectrum in this range has been modeled along simple lines. Qualitatively, the peak flux derived from the SN model as a function of redshift and shown here agrees with the theoretical peak flux-redshift relation for type Type Ia (Schmidt et al. 1998). This suggests that our adopted model for the UV spectrum is reasonable.

SECTION 7.C
Upper-limit Determination for Figure 7.1

In keeping with standard practice, our 17 April observations consisted of three separate 300-s observations (dithered by 5 arcseconds). Visual inspection of the three frames reveals a faint source near the position of the optical transient. In no frames did a diffraction spike of the nearby bright star "A" overlap the OT position. Also, there were no apparent cosmic-ray hits at the transient position nor were there any strong gain variations (that is, no apparent problem with the flat-fielding) at the three positions on the CCD.

In both the sum and mode-scaled median of the three shifted images, we detect a faint source consistent with the centroid location (angular offset 0.04 ± 0.18 arcsec) of the optical transient on 1998 March 27. Lastly, we computed the point source sensitivity in the 17 April image by computing Gaussian-weighted photometry in 1000 random apertures (see discussion accompanying table 7.1). Relative to this distribution, the flux at the location of the transient is positive and equal to 4.6-σ; the probability that the measured flux is due to noise is 2×10^{-6}. All objects at the flux level of the transient are reliably detected in the deeper image from December, thereby providing an independent validation of our methodology. We conclude that indeed the transient was significantly detected on 17 April. We discuss the photometric calibration of the detection in table 7.1.

SECTION 7.D
Reduction Details for Figure 7.3

7.D.1 Observing details

Spectroscopic observations of the OT were obtained on 29 March 1998 UT, using the Low Resolution Imaging Spectrometer (LRIS; Oke et al. 1995) at the Keck-II 10-m telescope on Mauna Kea, Hawaii. We used a grating with 300 lines mm^{-1} blazed at $\lambda_{\text{blaze}} \approx 5000$ Å and a 1.0 arcsec wide slit. The effective wavelength coverage was $\lambda \sim 4000 - 9000$ Å and the instrumental resolution was ~ 12 Å. Two exposures of 1800 s each were obtained. We used Feige 34 (Massey et al. 1988) for flux calibration. The estimated uncertainty of the flux zero point is about 20%. Additional spectra were obtained on 23 April 1998 UT, in photometric conditions, using the same instrument, except that the spectrograph slit was 1.5 arcsec wide. The effective spectral resolution for this observations was ~ 16 Å. Three exposures of 1800 s each were obtained. For these observations we used HD 84937 (Oke & Gunn 1983) for flux calibration. The estimated zero-point uncertainty is about 10%. On both epochs, exposures of arc lamps were used for primary wavelength calibration. Night sky lines were used to correct for calibration changes due to flexure. In both cases, slit position angles were close to the parallactic angles. Thus the differential slit losses were negligible.

The spectra shown were convolved with a Gaussian with $\sigma = 5$ Å (that is, less than the instrumental resolution) and re-binned to a common 5 Å sampling. None of the apparent features in the spectra are real, on the basis of a careful examination of two-dimensional, sky-subtracted spectroscopic images: apparent emission of absorption features are all due to an imperfect sky subtraction noise. A sky spectrum from the April 23 observation, extracted in the same aperture, is shown for the comparison. These spectra are shown before the correction for the Galactic foreground extinction.

7.D.2 Spectrophotometry

In both epoch, we chose a slit position angle close to parallactic so that the slit would cover both the transient and a relatively bright star ($R \sim 19$ mag). The spectroscopic R-band magnitudes reported in table 7.1 were derived relative to the calibrated R-band magnitude of these stars. This calibration serves to eliminate most of the systematics and calibration errors; that is, the spectrophotometric magnitudes were put on the direct CCD system, and are not based on the flux calibration of the spectra (which do, nevertheless, agree to 20 percent). This procedure bypasses most of the systematic errors in comparing our spectroscopic magnitudes with those from direct CCD images.

113

CHAPTER 8

Detection of a supernova signature associated with GRB 011121[†]

J. S. Bloom[1], S. R. Kulkarni[1], P. A. Price[1,2], D. Reichart[1], T. J. Galama[1], B. P. Schmidt[2], D. A. Frail[1,3], E. Berger[1], P. J. McCarthy[8], R. A. Chevalier[4], J. C. Wheeler[5], J. P. Halpern[6], D. W. Fox[1], S. G. Djorgovski[1], F. A. Harrison[1], R. Sari[7], T. S. Axelrod[2], R. A. Kimble[9], J. Holtzman[10], K. Hurley[11], F. Frontera[12,13], L. Piro[14], & E. Costa[14]

1 Division of Physics, Mathematics and Astronomy, 105-24, California Institute of Technology, Pasadena, CA 91125

2 Research School of Astronomy & Astrophysics, Mount Stromlo Observatory, via Cotter Rd., Weston Creek 2611, Australia

3 National Radio Astronomy Observatory, Socorro, NM 87801

4 Department of Astronomy, University of Virginia, P.O. Box 3818, Charlottesville, VA 22903-0818

5 Astronomy Department, University of Texas, Austin, TX 78712

6 Columbia Astrophysics Laboratory, Columbia University, 550 West 120th Street, New York, NY 10027

7 Theoretical Astrophysics 130-33, California Institute of Technology, Pasadena, CA 91125

8 Carnegie Observatories, 813 Santa Barbara Street, Pasadena, CA 91101

9 Laboratory for Astronomy and Solar Physics, NASA Goddard Space Flight Center, Code 681, Greenbelt, MD 20771

10 Department of Astronomy, MSC 4500, New Mexico State University, P.O. Box 30001, Las Cruces, NM 88003

11 University of California at Berkeley, Space Sciences Laboratory, Berkeley, CA 94720-7450

12 Istituto Astrofisica Spaziale e Fisica Cosmica, C.N.R., Via Gobetti, 101, 40129 Bologna, Italy

13 Physics Department, University of Ferrara, Via Paradiso, 12, 44100 Ferrara, Italy

14 Istituto Astrofisica Spaziale, C.N.R., Area di Tor Vergata, Via Fosso del Cavaliere 100, 00133 Roma, Italy

[†] A version of this chapter was published in the *The Astrophysical Journal Letters*, vol. 572, L45–L49.

Abstract

Using observations from an extensive monitoring campaign with the *Hubble Space Telescope*, we present the detection of an intermediate-time flux excess that is redder in color relative to the afterglow of GRB 011121, currently distinguished as the gamma-ray burst with the lowest known redshift. The red "bump," which exhibits a spectral roll-over at ~ 7200 Å, is well described by a redshifted Type Ic supernova that occurred approximately at the same time as the gamma-ray burst event. The inferred luminosity is about half that of the bright supernova 1998bw. These results serve as compelling evidence for a massive star origin of long-duration gamma-ray bursts. Models that posit a supernova explosion weeks to months preceding the gamma-ray burst event are excluded by these observations. Finally, we discuss the relationship between spherical core-collapse supernovae and gamma-ray bursts.

SECTION 8.1
Introduction

Two broad classes of long-duration gamma-ray burst (GRB) progenitors have survived scrutiny in the afterglow era: the coalescence of compact binaries (see Fryer et al. 1999a for review) and massive stars (Woosley 1993). More exotic explanations (e.g., Paczyński 1988; Carter 1992; Dermer 1996) fail to reproduce the observed redshift distribution, detection of transient X-ray lines, and/or the distribution of GRBs about host galaxies.

In the latter viable scenario, the so-called "collapsar" model (Woosley 1993; MacFadyen & Woosley 1999; Hansen 1999), the core of a massive star collapses to a compact stellar object (such as a black hole or magnetar) which then powers the GRB while the rest of the star explodes. We expect to see two unique signatures in this scenario: a rich circumburst medium fed by the mass-loss wind of the progenitor (Chevalier & Li 1999) and an underlying supernova (SN). Despite extensive broadband modeling of afterglows, unambiguous signatures for a wind-stratified circumburst media have not been seen (e.g., Frail et al. 2000c; Berger et al. 2001b).

There has, however, been been tantalizing evidence for an underlying SN. The first association of a cosmologically distant GRB with the death of a massive star was found for GRB 980326, where a clear excess of emission was observed, over and above the rapidly decaying afterglow component. This late-time "bump" was interpreted as arising from an underlying SN (Bloom et al. 1999c) since, unlike the afterglow, the bump was very red. GRB 970228, also with an intermediate-time bump and characteristic SN spectral rollover, is another good candidate (Reichart 1999; Galama et al. 2000).

Suggestions of intermediate-time bumps in GRB light curves have since been put forth for a number of other GRBs (Lazzati et al. 2001; Sahu et al. 2000; Fruchter et al. 2000c; Björnsson et al. 2001; Castro-Tirado et al. 2001; Sokolov 2001; Dar & Rújula 2002). Most of these results are tentative or suspect with the SN inferences relying on a few mildly deviant photometric points in the afterglow light curve. Even if some of the bumps are real, a number of other explanations for the physical origin of such bump have been advanced: for example, dust echoes (Esin & Blandford 2000; Reichart 2001), shock interaction with circumburst density discontinuities (e.g., Ramirez-Ruiz et al. 2001), and thermal re-emission of the afterglow light (Waxman & Draine 2000). To definitively distinguish between the SN hypothesis and these alternatives, detailed spectroscopic and multi-color light curve observations of intermediate-time bumps are required.

It is against this background that we initiated a program with the *Hubble Space Telescope* (HST) to sample afterglow light curves at intermediate and late-times. The principal attractions of HST are the photometric stability and high angular resolution. These are essential in separating the afterglow from the host galaxy and in reconstructing afterglow colors.

On theoretical grounds, if the collapsar picture is true, then we expect to see a Type Ib/Ic SN (Woosley 1993). In the first month, core-collapsed supernova spectra are essentially characterized by a blackbody (with a spectral peak near ~ 5000 Å) modified by broad metal-line absorption and a strong flux suppression blueward of ~ 4000 Å in the restframe. For GRBs with low redshifts, $z \lesssim 1$, the effect of this blue absorption blanketing is a source with an apparent red spectrum at observer-frame optical wavelengths; at higher redshifts, any supernova signature is highly suppressed. For low redshift GRBs, intermediate-time follow-up are, then, amenable to observations with the Wide Field Planetary Camera 2 (WFPC2). In this *Letter* we report on WFPC2 multi-color photometry of GRB 011121 ($z = 0.36$; Infante et al. 2001) and elsewhere we report on observations of GRB 010921 ($z = 0.451$; Price et al. 2002a). In a companion paper (Price et al. 2002; hereafter Paper II), we report a multi-wavelength (radio, optical and NIR) modeling of the afterglow.

8.2 Observations and Reductions

8.2.1 Detection of GRB 011121 and the afterglow

On 21.7828 November 2001 UT, the bright GRB 011121 was detected and localized by *BeppoSAX* to a 5-arcmin radius uncertainty (Piro et al. 2001a). Subsequent observations of the error circle refined by the IPN and *BeppoSAX* (see Paper II) revealed a fading optical transient (OT) (Wyrzykowski et al. 2001; Stanek et al. 2001). Spectroscopic observations with the Magellan 6.5-m telescope revealed redshifted emission lines at the OT position ($z = 0.36$), indicative of a bright, star-forming host galaxy of GRB 011121 (Infante et al. 2001).

8.2.2 HST Observations and reductions

For all the HST visits, the OT and its underlying host were placed near the serial readout register of WF chip 3 (position WFALL) to minimize the effect of charge transfer (in)efficiency (CTE). The data were pre-processed with the best bias, dark, and flat-field calibrations available at the time of retrieval from the archive ("on–the–fly" calibration). We combined all of the images in each filter, dithered by sub-pixel offsets, using the standard IRAF/DITHER2 package to remove cosmic rays and produce a better sampled final image in each filter. An image of the region surrounding the transient is shown in figure 8.1. The point source was detected at better than 20 σ in epochs one, two and three in all filters, and better than 5 σ in epoch four.

Given the proximity of the OT to its host galaxy, the final HST images were photometered using the IRAF/DAOPHOT package which implements PSF-fitting photometry on point-sources (Stetson 1987). The PSF local to the OT was modeled with **PSTSELECT** and **PSF** using at least 15 isolated stars detected in the WF chip 3 with an adaptive kernel to account for PSF variations across the image (VARORDER = 1). The resulting photometry, reported in Table 8.1, was obtained by finding the flux in an $0''.5$ radius using a PSF fit. We corrected the observed countrate using the formulation for CTE correction in Dolphin (2000) with the most up-to-date parameters[1]; such corrections, computed for each individual exposure, were never larger than 8% (typically 4%) for a final drizzled image. We estimated the uncertainty in the CTE correction, which is dependent upon source flux, sky background, and chip position, by computing the scatter in the CTE corrections for each of the images that were used to produce the final image. The magnitudes reported in the standard bandpass filters in Table 8.1 were found using the Dolphin prescription.

8.3 Results

In figure 8.2 we plot the measured fluxes from our four HST epochs in the F555W, F702W, F814W and F850LP filters. We also plot measurements made at earlier times (0.5 days $< t <$ 3 days) with

---
[1] See http://www.noao.edu/staff/dolphin/wfpc2_calib/.

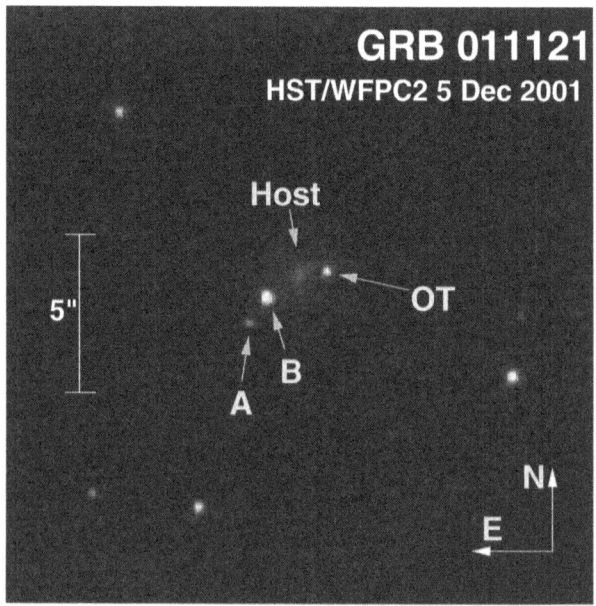

Figure 8.1 *Hubble Space Telescope* image of the field of GRB 011121 on 4–6 December 2001 UT. This false-color image was constructed by registering the final drizzled images in the F555W (blue), F702W (green) and F814W (red) filters. The optical transient (OT) is clearly resolved from the host galaxy and resides in the outskirts of the morphologically smooth host galaxy. Following the astrometric methodology outlined in Bloom et al. (2002), we find that the transient is offset from the host galaxy (883 ± 7) mas west, (86 ± 13) mas north. The projected offset is (4.805 ± 0.035) kpc, almost exactly at the host half-light radius. Sources "A" and "B" are non-variable point sources that appear more red than the OT and are thus probably foreground stars.

Table 8.1. Log of HST Imaging and Photometry of the OT of GRB 011121

Filter	Δt^a (days)	Integration Time (sec)	λ_{eff} (Å)	$f_\nu(\lambda_{\text{eff}})$ (μJy)	Vega Magnitude[b] (mag)
			Epoch 1		
F450W	13.09	1600	4678.52	0.551 ± 0.037	$B = 24.867 \pm 0.073$
F555W	13.16	1600	5560.05	0.996 ± 0.049	$V = 23.871 \pm 0.056$
F702W	13.23	1600	7042.48	1.522 ± 0.072	$R = 23.211 \pm 0.054$
F814W	14.02	1600	8110.44	1.793 ± 0.042	$I = 22.772 \pm 0.032$
F850LP	14.15	1600	9159.21	1.975 ± 0.103	
			Epoch 2		
F555W ...	23.03	1600	5630.50	0.647 ± 0.035	$V = 24.400 \pm 0.061$
F702W ...	23.09	1600	7002.71	1.271 ± 0.051	$R = 23.382 \pm 0.048$
F814W ...	24.83	1600	8105.05	1.495 ± 0.053	$I = 22.982 \pm 0.043$
F850LP ...	24.96	1600	9166.39	1.708 ± 0.100	
			Epoch 3		
F555W ...	27.24	1600	5711.00	0.378 ± 0.027	$V = 25.071 \pm 0.076$
F702W ...	27.30	1600	7043.85	0.981 ± 0.036	$R = 23.697 \pm 0.044$
F814W ...	28.10	1600	8164.90	1.301 ± 0.070	$I = 23.157 \pm 0.061$
F850LP ...	28.16	1600	9188.39	1.635 ± 0.092	
			Epoch 4		
F555W ...	77.33	2100	5604.61	0.123 ± 0.014	$V = 26.173 \pm 0.118$
F702W ...	76.58	4100	7042.09	0.224 ± 0.019	$R = 25.264 \pm 0.092$
F814W ...	77.25	2000	8149.18	0.294 ± 0.020	$I = 24.762 \pm 0.073$

Note. — In the fourth column, the effective wavelength of the filter based upon the observed spectral flux distribution of the transient at the given epoch. In the fifth column, the flux is given at this effective wavelength in an $0''.5$ radius. The observed count rate, corrected for CTE effects, was converted to flux using the IRAF/SYNPHOT package. An input spectrum with f_ν = constant was first assumed. Then approximate spectral indices between each filter were computed and then used to re-compute the flux and the effective wavelength of the filters. This bootstrapping converged after a few iterations. The HST photometry contains an unknown but small contribution from the host galaxy at the OT location. We attempted to estimate the contamination of the host at the transient position by measuring the host flux in several apertures at approximate isophotal levels to the OT position. We estimate the contribution of the host galaxy to be $f_\nu(F450W) = (0.098 \pm 0.039)$ μJy, $f_\nu(F555W) = (0.087 \pm 0.027)$ μJy, $f_\nu(F702W) = (0.127 \pm 0.026)$ μJy, $f_\nu(F814W) = (0.209 \pm 0.059)$ μJy, and $f_\nu(F850LP) = (0.444 \pm 0.103)$ μJy. To correct these numbers to "infinite aperture," multiply the fluxes by 1.096 (Holtzman et al. 1995). These fluxes have not been corrected for Galactic or host extinction.

[a] Mean time since GRB trigger on 21.7828 Nov 2001 UT.

[b] Tabulated brightnesses in the Vega magnitude system ($B_{\text{Vega}} = 0.02$ mag, $V_{\text{Vega}} = 0.03$ mag, $R_{\text{Vega}} = 0.039$ mag, $I_{\text{Vega}} = 0.035$ mag; Holtzman et al. 1995). Subtract 0.1 mag from these values to get the infinite aperture brightness. These magnitudes have not been corrected for Galactic or host extinction.

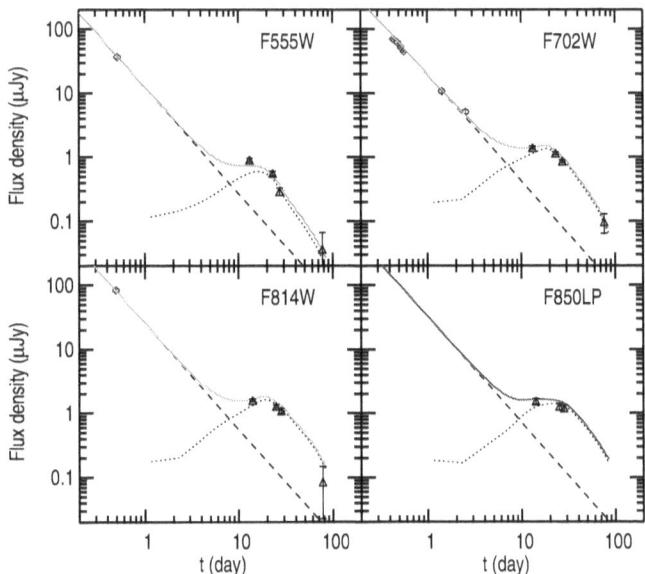

Figure 8.2 Light-curves of the afterglow and the intermediate-time red bump of GRB 011121. The triangles are our HST photometry in the F555W, F702W, F814W and F850LP filters (all corrected for the estimated contribution from the host galaxy), and the diamonds are ground-based measurements from the literature (Olsen et al. 2001; Stanek & Wyrzykowski 2001). The dashed line is our fit to the optical afterglow (see Paper II), the dotted line is the expected flux from the template SN at the redshift of GRB 011121, with foreground extinction applied and dimmed by 55% to approximately fit the data, and the solid line is the sum of the afterglow and SN components. Corrections for color effects between the ground-based filters and the HST filters were taken to be negligible for the purpose of this exercise.

ground-based telescopes and reported in the literature. These magnitudes were converted to fluxes using the zero-points of Fukugita et al. (1995) and plotted in the appropriate HST filters.

Corrections for color effects between the ground-based filters and HST filters were taken to be negligible for the purpose of this exercise.

The estimated contribution from the afterglow is heavily weighted by the available data: our ground-based data (and those reported in the literature so far) are primarily at early times. Roughly, over the first week, the afterglow exhibits a simple power law decay. The afterglow contribution derived from our NIR data and optical data from the literature (see Paper II) is shown by the dashed line in each panel. No afterglow light curve breaks (e.g., from jetting) were assumed.

Garnavich et al. (2002) drew attention to an excess of flux (in R-band), at a time 13 days after the GRB, with respect to that expected from the power-law extrapolation of early-time afterglow emission; they suggested the excess to arise from an underlying SN. As can vividly be seen from our multi-color data, the excess is seen in all bands and over several epochs.

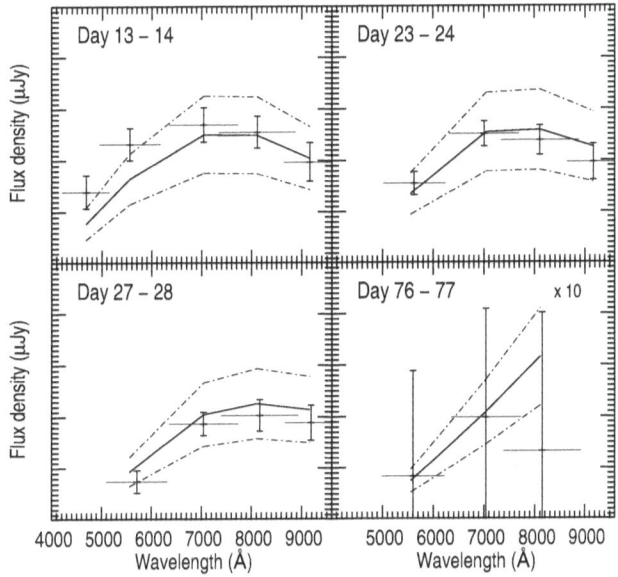

Figure 8.3 The spectral flux distributions of the red bump at the time of the four HST epochs. The fluxes are dereddened using $A_V = 1.16$ mag. Spectral evolution, and more important, a turn-over in the spectra of the first three epochs, are clearly seen. The peak of the turn-over (around 7200 Å) corresponds to a peak in the red bump spectrum at ~5300 Å. For comparison, we show a template broadband SN spectra (a dimmed version of SN 1998bw; solid curve) as it would appear at the redshift of GRB 011121 and the associated 2 σ errors (see text). The vertical error bars on the red bump reflect the 1 σ statistical uncertainty flux from only the red bump. There are large (~1 mag) systematic uncertainties (e.g., Galactic reddening, relative distance moduli between SN 1998bw and GRB 011121) in both the data and the model; these are suppressed for clarity.

We used the light curve and spectra[2] of the well-studied Type Ic supernova SN 1998bw (Galama et al. 1998b; McKenzie & Schaefer 1999) to create a comparison template broad-band light curve of a Type Ic supernova at redshift $z = 0.36$. Specifically, the spectra of SN 1998bw were used to compute the K-corrections between observed photometric bands of 1998bw and HST bandpasses (following Kim et al. 1996 and Schmidt et al. 1998). A flat Λ cosmology with $H_0 = 65$ km s^{-1} Mpc^{-1} and $\Omega_M = 0.3$ was assumed and we took the Galactic foreground extinction to SN 1998bw of $A_V = 0.19$ mag (Galama et al. 1998b).

Since dimmer Ic SNe tend to peak earlier and decay more quickly (see fig. 1 of Iwamoto et al. 1998), much in the same way that SN Ia do, we coupled the flux scaling of SN 1998bw with time scaling in a method analogous to the "stretch" method for SN Ia distances (Perlmutter et al. 1997). To do so, we fit an empirical relation between 1998bw and 1994I to determine the flux-time scaling. We estimate that a 1998bw-like SN that is dimmed by 55% (see below), would peak and decay about 17% faster than 1998bw itself. Some deviations from our simple one-parameter template are apparent, particularly in the F555W band and at late-times.

In figure 8.3, we plot the spectral flux distributions (SFDs) of the intermediate-time bump at the four HST epochs. A clear turn-over in the spectra in the first 3 epochs is seen at about 7200 Å. The solid curve is the SFD of SN 1998bw transformed as described above with the associated 2-σ errors. Bearing in mind that there are large systematic uncertainties in the template (i.e., the relative distance moduli between SN 1998bw and GRB 011121) and in the re-construction of the red bump itself (i.e., the Galactic extinction toward GRB 011121 and the contribution from the afterglow in the early epochs), the consistency between the measurements and the SN is reasonable. We consider the differences, particularly the bluer bands in epoch one, to be relatively minor compared with the overall agreement. This statement is made in light of the large observed spectral diversity of Type Ib/Ic SNe (see, for example, figure 1 of Mazzali et al. 2002).

SECTION 8.4

Discussion and Conclusions

We have presented unambiguous evidence for a red, transient excess above the extrapolated light curve of the afterglow of GRB 011121. We suggest that the light curve and spectral flux distribution of this excess appears to be well represented by a bright SN. While we have not yet explicitly compared the observations to the expectations of alternative suggestions for the source of emission (dust echoes, thermal re-emission from dust, etc.), the simplicity of the SN interpretation—requiring only a (physically motivated) adjustment in brightness—is a compelling (i.e., Occam's Razor) argument to accept our hypothesis. Given the red bump detections in a number of other GRBs occur on a similar timescale as in GRB 011121, any model for these red bumps should have a natural timescale for peak of $\sim 20(1+z)$ day; in our opinion, the other known possibilities do not have such a natural timescale as compared with the SN hypothesis. Indeed, if our SN hypothesis is correct, then the flux should decline as an exponential from epoch four onward. The ultimate confirmation of the supernova hypothesis is a spectrum which should show characteristic broad metal-line absorption of the expanding ejecta (from, e.g., Ca II, Ti II, Fe II).

We used a simplistic empirical brightness–time stretch relation to transform 1998bw, showing good agreement between the observations and the data. If we neglect the time-stretching and only dim the 1998bw template, then the data also appear to match the template reasonably well, however, the discrepancies in the bluer bands become somewhat larger and the flux ratios between epochs are slightly more mismatched. The agreement improves if we shift the time of the supernova to be about $\sim$3–5 days (restframe) before the GRB time. Occurrence times more than about ten days (restframe) before the GRB can be ruled out. This observation, then, excludes the original "supranova" idea (Vietri &

---
[2] Spectra were obtained through the Online Supernova Spectrum Archive (SUSPECT) at http://tor.nhn.ou.edu/~suspect/index.html.

Stella 1998), that posited a supernova would precede a GRB by several years (see eq. [1] of Vietri & Stella 1998). Modified supranova scenarios that would allow for any time delay between the GRB and the accompanying SN, albeit *ad hoc*, are still consistent with the data presented herein[3]

Regardless of the timing between the SN explosion and the GRB event (constrained to be less than about 10 days apart), the bigger picture we advocate is that GRB 011121 resulted from an explosive death of a massive star. This conclusion is independently supported by the inference, from afterglow observations of GRB 011121 (Paper II), of a wind-stratified circumburst medium.

The next phase of inquiry is to understand the details of the explosion and also to pin down the progenitor population. A large diversity in any accompanying SN component of GRBs is expected from both a consideration of SNe themselves and the explosion mechanism. The three main physical parameters of a Type Ib/Ic SN are the total explosive energy, the mass of the ejecta, and the amount of Nickel synthesized by the explosion (M_{Ni}). The peak luminosity and time to peak are roughly determined by the first two whereas the exponential tail is related to M_{Ni}. Ordinary Ib/Ic SNe appear to show a wide dispersion in the peak luminosity (Iwamoto et al. 1998). There is little *ab initio* understanding of this diversity (other than shifting the blame to dispersion in the three parameters discussed above).

It is now generally accepted that GRBs are not spherical explosions and are, as such, usually modeled as a jetted outflow. Frail et al. (2001) model the afterglow of GRBs and have presented a compilation of opening angles, θ, ranging from less than a degree to 30 degrees and a median of 4 degrees. If GRBs have such strong collimation then it is not reasonable to assume that the explosion, which explodes the star, will be spherical. We must be prepared to accept that the SN explosion is extremely asymmetric and thus even a richer diversity in the light curves. This expected diversity may account for both the scale factor difference between the SN component seen here and in SN 1998bw seen in figure 8.3. Indeed, there has been a significant discussion as to the degree to which the central engine in GRBs will affect the overall explosion of the star (Woosley 1993; Khokhlov et al. 1999; MacFadyen & Woosley 1999; Höflich et al. 1999; MacFadyen et al. 2001). These models have focused primarily on the hydrodynamics and lack the radiative modeling necessary to compare observations to the models.

Clearly, the observational next step is to obtain spectroscopy (and perhaps even spectropolarimetry) and to use observations to obtain a rough measure of the three-dimensional velocity field and geometry of the debris. As shown by GRB 011121 the SN component is bright enough to undertake observations with the largest ground-based telescopes.

We end by noting the following curious point. The total energy yield of a GRB is usually estimated from the gamma-ray fluence and an estimate of θ (see Frail et al. 2001). Alternatively, the energy in the afterglow is used (e.g., Piran et al. 2001). However, for GRB 011121, the energy in the SN component (scaling from the well-studied SN 1998bw) is likely to be comparable or even larger than that seen in the burst or the afterglow. In view of this, the apparent constancy of the γ-ray energy release is even more mysterious.

We thank S. Woosley, who, as referee, provided helpful insights toward the improvement of this work. A. MacFadyen and E. Ramirez-Ruiz are acknowledged for their constructive comments on the paper. J. S. B. is a Fannie and John Hertz Foundation Fellow. F. A. H. acknowledges support from a Presidential Early Career award. S. R. K. and S. G. D. thank the NSF for support. R. S. is grateful for support from a NASA ATP grant. R. S. and T. J. G. acknowledge support from the Sherman Fairchild Foundation. J. C. W. acknowledges support from NASA grant NAG59302. KH is grateful for Ulysses support under JPL Contract 958056, and for IPN support under NASA Grants FDNAG 5-11451 and NAG 5-17100. Support for Proposal number HST-GO-09180.01-A was provided by NASA through a grant from Space Telescope Science Institute, which is operated by the Association of Universities for Research in Astronomy, Incorporated, under NASA Contract NAS5-26555.

---

[3] The explosion date of even very well-studied supernovae, such as 1998bw, cannot be determined via light curves to better than about 3 days (e.g., Iwamoto et al. 1998). This implies that future photometric studies might not be equipped to distinguish between contemporaneous SN/GRB events and small delay scenarios.

CHAPTER 9

Expected Characteristics of the Subclass of Supernova Gamma-Ray Bursts[†]

J. S. Bloom[1], S. R. Kulkarni[1], F. Harrison[1], T. Prince[1], E. S. Phinney[1], D. A. Frail[2]

[1] Palomar Observatory 105-24, California Institute of Technology, Pasadena, CA 91125, USA
[2] National Radio Astronomy Observatory, P.O. Box O, 1003 Lopezville Road, Socorro, NM

Abstract

The spatial and temporal coincidence of gamma-ray burst (GRB) 980425 and supernova (SN) 1998bw has prompted speculation that there exists a subclass of GRBs produced by SNe ("S-GRBs"). A physical model motivated by radio observations lead us to propose the following characteristics of S-GRBs: (1) prompt radio emission and an implied high brightness temperature close to the inverse Compton limit, (2) high expansion velocity ($\gtrsim$50,000 km s-1) of the optical photosphere as derived from lines widths and energy release larger than usual, (3) no long-lived X-ray afterglow, and (4) a single-pulse GRB profile. Radio studies of previous SNe show that only (but not all) Type Ib and Ic SNe potentially satisfy the first condition. We investigate the proposed associations of GRBs and SNe within the context of these proposed criteria and suggest that $\sim$1% of GRBs detected by BATSE may be members of this subclass.

SECTION 9.1

Introduction

With the spectroscopic observations of the optical afterglow of gamma-ray burst (GRB) 970508 by Metzger et al. (1997b) came proof that at least one GRB is at a cosmological distance. Kulkarni et al. (1998b) later added another cosmological GRB, which, based on an association with a high-redshift galaxy, had an implied energy release of $E_\gamma \gtrsim 10^{53}$ erg. However, not all GRBs have been shown to be associated with distant host galaxies. Only about half of all GRBs are followed by long-lived optical afterglow, and one in four produce a longer-lived radio afterglow at or above the 100 μJy level. In contrast, X ray afterglow has been seen for almost all *BeppoSAX*-localized bursts. Until recently, the emerging picture had been that all GRBs are located at cosmological distances and these GRBs (hereafter cosmological GRBs, or C-GRBs) are associated with star-forming regions and that C-GRBs are the death throes of massive stars.

The discovery of a supernova (SN 1998bw) both spatially (chance probability of 10^{-4}) and temporally coincident with GRB 980425 (Galama et al. 1998; Galama et al. 1998b) suggests the existence of another

[†] A version of this chapter was first published in *The Astrophysical Journal Letters*, 506, p. L105–L108, (1998).

class of GRBs. Remarkably, SN 1998bw showed very strong radio emission with rapid turn-on; it is, in fact, the brightest radio SN to date (Wieringa et al. 1998). This rarity further diminishes the probability of chance coincidence (Sadler et al. 1998). From the radio observations, Kulkarni et al. (1998) concluded that there exists a relativistic shock [bulk Lorentz factor, $(\Gamma \equiv (1 - \beta^2)^{-1/2} \gtrsim 2]$ even 4 days after the SN explosion. Kulkarni et al. argue that the young shock had all the necessary ingredients (high Γ, sufficient energy) to generate the observed burst of gamma rays.

We feel that the physical connection between GRB 980425 and SN 1998bw is strong. Accepting this connection then implies that there is at least one GRB that is not of distant cosmological origin but is instead related to an SN event in the local universe ($\lesssim 100$ Mpc). We refer to this category of GRBs as supernova-GRBs or S-GRBs. Many questions arise: How common are S-GRBs? How can they be distinguished from C-GRBs? What are their typical energetics?

In this paper, accepting the physical model advocated by Kulkarni et al. (1998), we enumerate the defining characteristics of the class of S-GRBs. We then apply these criteria to members of this proposed class and conclude with a discussion of the potential number of S-GRBs.

SECTION 9.2
How to Recognize S-GRBs

The expected characteristics of S-GRBs is motivated by the model developed to explain the radio observations of SN 1998bw. Briefly, from the radio data, Kulkarni et al. (1998) conclude that the radio emitting region is expanding at least at $2c$ (4 days after the explosion) and slowing down to c, one month after the burst. Indeed, one expects the shock to slow down as it accretes ambient matter. Thus, it is reasonable to expect the shock to have had a higher Γ when it was younger. The expectation is that this high-Γ shock is also responsible for the observed burst of gamma rays (synchrotron or inverse Compton scattering). Of note, whereas in C-GRBs the primary afterglow is optical, in S-GRBs the primary afterglow is in the radio band. We now enumerate the four criteria of S-GRBs:

1. *Prompt radio emission and high brightness temperature.*—An unambiguous indication of a relativistic shock in an SN is when the inferred brightness temperature, T_B, exceeds $T_{\rm icc} \sim 4 \times 10^{11}$ K, the so-called inverse Compton catastrophe temperature. T_B is given by

$$T_B = 6 \times 10^8 \Gamma^{-3} \beta^{-2} S({\rm mJy})(\nu/5\,{\rm GHz})^{-2} t_d^{-2} d_{\rm Mpc}^2 \ {\rm K}; \tag{1}$$

here t_d is the time in days since the burst of gamma-rays, $d_{\rm Mpc}$ is the distance in Mpc, and S, the flux density at frequency ν. The energy in the particles and the magnetic field is the smallest when $T_B \simeq T_{\rm eq}$, the so-called "equipartition" temperature ($T_{\rm eq} \sim 5 \times 10^{10}$ K; Readhead 1994). The inferred energy increases sharply with increasing T_B. For SN 1998bw, even with $T_B = T_{\rm eq}$, the inferred energy in the relativistic shock is 10^{48} ergs, which is already significant. If $T_B > T_{\rm icc}$, the inferred energy goes up by a factor of 500 and thus approaches the total energy release of a typical SN ($\sim 10^{51}$ ergs). Thus, the condition $T_B < T_{\rm icc}$ is a reasonable inequality to use. This then leads to a lower limit on Γ. We consider the shock to be relativistic when $\Gamma\beta > 1$. For SN 1998bw, Kulkarni et al. (1998) find $\Gamma\beta \gtrsim 2$.

It is well known that prompt radio emission (by this we mean a timescale of a few days) is seen from Type Ib/Ic SNe (Weiler & Sramek 1998; Chevalier 1998). Radio emission in Type II SNe peaks on very long timescales (months to years). No Type Ia SN has yet been detected in the radio. Thus, the criterion of prompt radio emission (equivalent to high T_B) will naturally lead to selecting only Type Ib/Ic SNe. High brightness temperature is achieved when the radio flux is high. Indeed, the radio luminosity of SN 1998bw was 2 orders of magnitude larger than the five previously studied Type Ic/Ic SNe (van Dyk et al. 1993).

2. *No long-lived X-ray afterglow.*—In our physical picture above, we do not expect any long-lived X-ray emission since the synchrotron lifetime of X-ray–emitting electrons is so short. The lack of X-ray afterglow from GRB 980425 in the direction of SN 1998bw is consistent with this picture.

3. *A simple GRB profile.*—In the model we have adopted, the gamma-ray and the radio emission is

powered by an energetic relativistic shock. Is it likely that there is more than one relativistic shock? Our answer is no. There is no basis to believe or expect that the collapse of the progenitor core will result in multiple shocks. It is possible that the nascent pulsar or a black hole could be energetically important, but the envelope matter surely will dampen down rapid temporal variability of the underlying source. From this discussion we conclude that there is only one relativistic shock. Thus, the gamma-ray burst profile should be very simple: a single pulse (SP).

The light curve of GRB 980425 (fig. 9.1) is a simple single pulse (SP) with a $\sim$5 s rise (HWHM) and a $\sim$8 s decay. Like most GRBs (e.g., Crider et al. 1997; Band 1997), the harder emission precedes the softer emission with channel 3 (100–300 keV) peaking $\sim$1 s before channel 1 (25–50 keV). Unlike most GRB light curves, the profile of GRB 980425 has a rounded maximum instead of a cusp.

4. *Broad line emission and bright optical luminosity.*—Kulkarni et al. (1998) noted that the minimum energy in the relativistic shock, E_{min}, is 10^{48} ergs and that the true energy content could be as high as 10^{52} ergs. Even the lower value is a significant fraction of energy of the total supernova release of ordinary SNe ($E_{tot} \sim 10^{51}$ ergs). Clearly, a larger energy release in the supernova would favor a more energetic shock and, hence, increase the chance such a shock could produce a burst of gamma rays. Indeed, there are indications from the modeling of the light curve and the spectra that the energy release in SN 1998bw was 3×10^{52} ergs (Woosley et al. 1999; Iwamoto et al. 1998), a factor of $\sim$30 larger than the canonical SN. This then leads us to propose the final criterion: indications of a more-than-normal release of energy. Observationally, this release is manifested by large expansion speed, which leads to the criterion of broad emission lines and bright optical luminosity.

Nakamura (1998) suggests that S-GRBs derive their energy from the formation of a strongly magnetized pulsar rotating at millisecond period. Furthermore, he advocates that S-GRBs must possess "non-high-energy" (NHE; see Pendleton et al. 1997) profiles (i.e., little flux above 300 keV). However, Nakamura's model does not address the most outstanding feature of SN 1998bw—its extremely unusual radio emission. Our model is silent on whether the bursts should be NHE or HE since that would depend on the details of the emission mechanism and the importance of subsequent scattering.

There is an implicit assumption on the part of several authors (e.g., Nakamura 1998; Woosley et al. 1999; Wang & Wheeler 1998) that S-GRBs are intimately connected with Type Ic SNe. Within the framework of our model, the key issue is whether there exists a relativistic shock that can power the gamma rays. Clearly this relativistic shock is distinctly different from the low-velocity shock that powers the optical emission. Thus, the connection between GRB emission and the optical properties of the SN is bound to be indirect (e.g., our fourth criterion).

SECTION 9.3
Application of Criteria to Proposed Associations

We now apply the above four criteria motivated by a specific physical model to proposed S-GRBs (Wang & Wheeler 1998; Woosley et al. 1999). We searched for more potential GRB-SN associations by cross-correlating the earlier WATCH and Interplanetary Network (IPN) localizations (Atteia et al. 1987; Lund 1995; Hurley et al. 1997) with an archive catalog of supernovae[1]. We found no convincing associations in archival GRB/SN data before the launch of the Burst and Source Transient Experiment (BATSE). Thus, our total list remains at nine, seven from Wang & Wheeler (1998) and two from Woosley et al. (1999).

We reject the following proposed associations: (1) SN 1996N/GRB 960221 (Wang & Wheeler 1998). The IPN data rule out this association on spatial grounds alone. This lack of association was independently recognized by Kippen et al. (1998). (2) SN 1992ar/GRB 920616 (Woosley et al. 1999). The associated GRB appears not to exist in the BATSE 4B Catalog[2] and furthermore, there are no other

---

[1] The updated Asiago Supernova Catalog of Barbon et al. (1989) (maintained by E. Capellaro) is available at http://athena.pd.astro.it/~supern/.
[2] The BATSE Gamma-Ray Burst Catalog (maintained by C. A. Meegan et al.), including the BATSE 4B Catalog, is

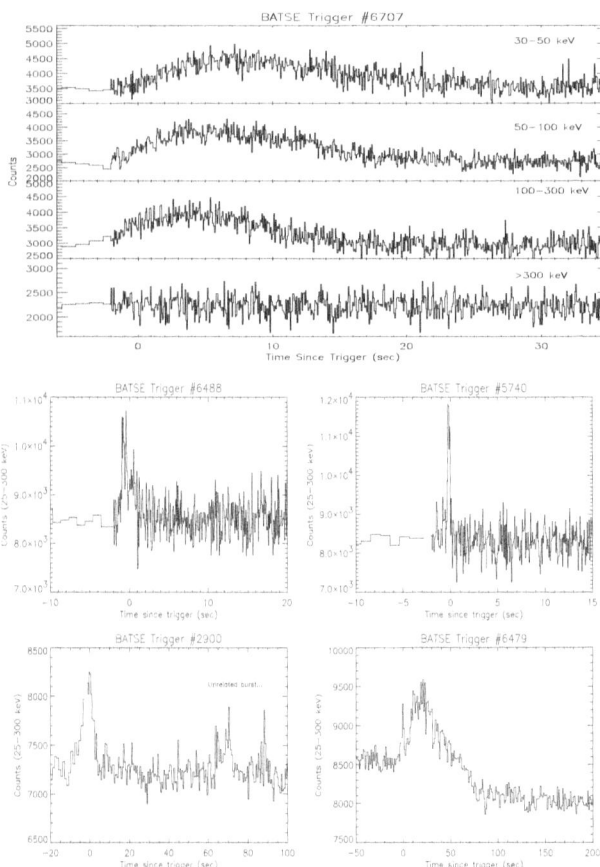

Figure 9.1 (top) The 4-channel light curve of GRB 980425 (Trigger #6707) associated with SN 1998bw. The single pulse (SP) appears cusp-less unlike most SP BATSE bursts. The hard-to-soft evolution is clear from the progression of the peak from channels 3 to 1 over time. After BATSE triggers, the light curve is sampled on 64-ms timescales. Continuous DISCLA data is augmented to the pre-trigger light curve; this data is basically a 16 bin (1.024 sec) averaged over the more finely sampled 64-ms data. In the case of longer bursts, we average the 64-ms bins over 16-sec intervals to reduce noise. (bottom panel; clockwise from top left) Light curve of GRB 97112 (Trigger #6488)/SN 1997ef or SN 1997ei; GRB 970103 (Trigger #5740)/SN 1997X; GRB 971115 (Trigger #6479)/SN 1997ef; GRB 940331 (Trigger #2900)/SN 1994I. According to the BATSE archive, the pulse beginning $t \simeq 65$ sec is an unrelated (i.e., not spatially coincident) GRB.

Table 9.1. GRB/Supernovae Associations: Which are Truly S-GRBs?

Association	SN Type	Prompt Radio?	$\delta\theta^{\clubsuit}$ ($N\sigma$)	v_{max} (km s^{-1})	GRB Type	$D^{\clubsuit}$ (Mpc)	$S^{\heartsuit}$ ($\times 10^{-7}$) (erg cm^{-2})	$E^{\diamond}$ (erg)
1998bw/6707	Ic	Y	0.0	60,000^a	SP/NHE	39.1^b	44^c	8.1 $\times 10^{47}$
1997ei/6488	Ic	NA	2.4	13,000^d	SP/NHE	48.9^e	7.69	2.2 $\times 10^{47}$
1997X/5740	Ic	NA	3.1	16,000^f	SP/HE	17.0^e	5.88	2.0 $\times 10^{46}$
1994I/2900	Ib/c	Y	4.4	14,000^g	SP/NHE	7.10^e	32.6	2.0 $\times 10^{46}$
1997ef/6488	Ib/c?	NA	NA	15,000^h	SP/NHE	53.8^e	7.69	2.7 $\times 10^{47}$
1997ef/6479	Ib/c?	NA	NA	15,000^h	MP/HE	53.8^e	99.5	3.4 $\times 10^{48}$
1992ad/1641	Ib	NC†	2.0	NA	NA	19.504^e	NA	NA
1997cy/6230	IIPec	NA	NA	5000^i	SP/HE	295^i	2.22	2 $\times 10^{48}$
1993J/2265	IIt	N	NA	13,000^j	SP/NHE	3.63^k	1.53^c	2.4 $\times 10^{44}$

Note. — GRB and SN properties of the suggested pairs by Wang & Wheeler (1998) and Woosley et al. (1999) are compared against the expected criteria of S-GRBs (see §9.2). The associations are listed in order of decreasing likelihood that the SN/GRB falls into the S-GRB subclass. Those with the least amount of information are placed at the bottom of the list. The list last two entries are SNe which are of type II and thus listed separately. NA = not available. ♣ Distance of SN from BATSE position in units of number of BATSE sigma from Kippen et al. (1998). In the case of GRB 980425, the SN 1998bw lies near the center of the small ($\sim$ 8 arcmin) BeppoSAX error circle. ♣ Assuming $H_0 = 65$ km s^{-1} Mpc^{-1} with $D \simeq cz/H_0$ with z, the heliocentric redshift, from noted reference. $^{\heartsuit}$ Fluence in BATSE channels 1 - 4 (24-1820 keV). From Meegan et al. (1998) (BATSE Database) unless noted. $^{\diamond}$ Required isotropic energy (> 25 keV). † The prompt radio criterion is not constrained (NC) by the late radio detections.

References. — ^{a} R. A. Stathakis communication in Kulkarni et al. (1998); ^{b} Tinney et al. (1998); ^{c} Galama et al. (1998), Galama et al. (1998b); ^{d} Based on a spectrum provided in Wang et al. (1998); ^{e} de Vaucouleurs et al. (1991); ^{f} Benetti et al. (1997b); ^{g} Wheeler et al. (1994); ^{h} Based on Filippenko (1997); ^{i} Benetti et al. (1997a); ^{j} Filippenko & Matheson (1993); ^{k} Freedman et al. (1994)

GRBs within a month that are spatially coincident with the SN. (3) SN 1998T/GRB 980218 (Wang & Wheeler 1998). This is ruled out on spatial grounds from the IPN data (Kippen et al. 1998).

In table 9.1 we summarize the proposed associations. They are ranked according to the viability of the association based on the four criteria discussed in the previous section. The pulse profile for each GRB is characterized as either simple/single pulse (SP) or multipulse (MP). The SN type was drawn from the literature, as was the distance to the host galaxy. The isotropic gamma-ray energy release is computed from the publicly available fluence (BATSE 4B Catalog) and the assumed distance.

It is unfortunate that crucial information—the early radio emission observations—are missing for all but one SN (1994I). SN 1994I does have early radio emission (Rupen et al. 1994). However, according to Kippen et al. (1998), the associated candidate GRB 940331 is more than 4 σ away from the location of SN 1994I. Thus, either the GRB associated with this event is not observed by BATSE, or this event is not an S-GRB.

SECTION 9.4
Discussion

From the observations (primarily radio) and analysis of SN 1998bw we have enumerated four criteria to identify S-GRBs. We have attempted to see how well the proposed associations of S-GRBs fare against these criteria. Unfortunately, we find the existing data are so sparse that we are unable to really judge if the proposed criteria are supported by the observations.

Independent of our four criteria, the expected rate of S-GRBs is constrained by the fact that this

available at http://www.batse.msfc.nasa.gov/data/grb/catalog/.

subclass is expected, with the assumption of a standard candle energy release, to have a homogeneous Euclidean ($\langle V/V_{\max}\rangle$=0.5) brightness distribution. Since there is a significant deviation from Euclidean in the BATSE catalog (e.g., Fenimore et al. 1993), S-GRBs cannot comprise a majority fraction of the BATSE catalog. Indeed, as studies show (e.g., Pendleton et al. 1997), $\approx$ 25% of the BATSE GRB population can derive from homogeneous population.

From the *BeppoSAX* observations we know that at least 90% of *BeppoSAX* -identified GRBs have an X-ray afterglow. Thus, at least in the *BeppoSAX* sample, the population of S-GRBs is further constrained to be no more than 10% using the criterion of no X-ray afterglow. However, it is well known that *BeppoSAX* does not trigger on short bursts—duration $\lesssim$ a few seconds—and thus this statement applies only to the longer bursts.

The small number of candidate associations prohibits us from drawing any firm conclusions based on common characteristics. Nonetheless, it is of some interest to note that four of our top five candidate S-GRBs (the exception is # 6479) are single-pulsed (SP) bursts. We clarify that the ordering in table 9.1 did not use the morphology of the pulse profile in arriving at the rank. We remind the reader that roughly half of all BATSE bursts are SP, and these mostly are sharp spikes (<1 s) or exhibit a fast rise followed by an exponential decay—the so-called FREDs. Thus, only a subclass of SP bursts could be S-GRBs.

What could be the special characteristics of this sub-class of SPs? In search of this special subclass, we note that the profile of GRB 980425 (fig. 9.1) exhibits a rounded maximum and is quite distinctive. A visual inspection of the BATSE 4B catalog shows that there are only 15 bursts with similar profiles; we note that such bursts constitute 1% of the BATSE bursts. Interestingly, most of these bursts appear to have the same duration as GRB 980425, although this may be due to bias in our selection. It is heartening to note that an independent detailed analysis of GRB light curves by Norris et al. (1999) confirmed the small fraction (1%-2%) of GRB light curves that meet our proposed criteria.

We end with some thoughts and speculation on the population of S-GRBs. Assuming the fluence of the GRB 980425 is indicative of the subclass, we find a canonical gamma-ray energy of $E \simeq 8 \times 10^{47}$ h_{65}^{-2} ergs. Although BATSE triggers on flux (rather than fluence), 80% of the bursts with fluence $S \gtrsim 8\times 10^{-7}$ ergs cm^{-2} will be detected (Bloom et al. 1996). Thus, BATSE can potentially probe the class of S-GRBs out to $\sim$100 h_{65}^{-1} Mpc. van den Bergh & Tammann (1991) concluded that the rate of Ib/Ic SNe is roughly half that of Type II SNe. Thus, the expected rate of Type Ib/Ic SNe is 0.3 per day out to a distance of 100 h_{65}^{-1} Mpc. This can be compared with the daily rate of $\sim$3 GRBs per day at the BATSE flux limit. Thus, if all Type Ib/Ic SNe produced an S-GRB, then the fraction of S-GRBs is 10%, consistent with the upper limit on the fraction due to the X-ray afterglow criterion found above. But, since most known SNe do not fit our criteria 1 and 4, the fraction constrained by the Type Ib/Ic rates is likely much smaller.

The sky distributions of SNe and GRBs that fit our four criteria but are not necessarily correlated (as in SN 1998bw/GRB 980425) can be used as an indirect test of the S-GRB hypothesis. Norris et al. (1999) have shown that the anisotropy of the 21 Type Ib/Ic SNe are marginally inconsistent with the isotropy of the 32 SP GRBs. We note, however, that most current search strategies are optimized to discover SNe in regions of large galaxy overdensity (presumably biased toward the supergalactic plane), which may cause the observed SNe anisotropy to be larger than it truly is. Further, most SNe Ib/Ic do not fit our criteria 1 and 4, and thus it is unwarranted to simply correlate all Type Ib/Ic SNe to SP GRBs.

We conclude with two suggestions for observations that directly test the S-GRB hypothesis. Even with the poor localization of BATSE, a Schmidt telescope equipped with large plates can be employed to search for SNe out to a few hundred megaparsecs. This is the best way to constrain the S-GRB population frequency. Second, S-GRBs will dominate the GRB number counts at the faint end of the flux distribution. From this perspective, future missions should be designed to have the highest sensitivity with adequate localization.

We thank J. Sievers, R. Simcoe, E. Waxman, B. Kirshner, A. Filippenko, S. Sigurdsson, and E. E. Fenimore for helpful discussions and direction at various stages of this work. The National Radio Astronomy Observatory is a facility of the National Science Foundation operated under cooperative agreement by Associated Universities, Inc. SRK and JSB are supported by the NSF and NASA.

Part III

An Instrument to Study the Small-scale Environments of GRBs

CHAPTER 10

JCAM: A Dual-Band Optical Imager for the Hale 200-inch Telescope at Palomar Observatory[†]

J. S. BLOOM[a], S. R. KULKARNI[a], J. C. CLEMENS[a,b], A. DIERCKS[a,c], R. A. SIMCOE[a], & B. B. BEHR[a,d]

[a] California Institute of Technology, MS 105-24, Pasadena, CA 91125 USA

[b] Department of Physics and Astronomy, University of North Carolina, Chapel Hill, NC 27599-3255, USA

[c] The Institute for Systems Biology, 1441 North 34th Street, Seattle, WA 98103-8904 USA

[d] RLM 15.308, C1400, UT-Austin, Austin, TX 78712 USA

Abstract

We describe the design and construction of the *Jacobs Camera* (JCAM), a dual-CCD optical imaging instrument now permanently mounted at the East Arm f/16 focus of the Hale 200 inch Telescope at Palomar Observatory. JCAM was designed to provide quick ($\lesssim$ 30 min) and ready access to high-quality photometry simultaneously in two optical bandpasses, albeit over a small field-of-view (3.2 arcmin diameter). The prime motivating science is as a follow-up imager to gamma-ray burst afterglows in the first hour to days after a burst. However, given the quick frame readout of each CCD (9.5 sec), JCAM may also be useful for time-resolved dual color photometry of other faint variables and as an effective tool for targeted surveys. JCAM, built for under $65 k, is the first instrument at Palomar to be fully operated remotely over the Internet.

SECTION 10.1
Introduction

The golden age of optical transient astronomy began in earnest over the past few years, fueled in part by the scientific promises of early gamma-ray burst (GRB) afterglow observations but also in part by the relative ease of constructing sophisticated instruments at low cost and mounting on inexpensive, dedicated telescopes. Robotic $\lesssim$ 0.5 meter-class instruments (e.g., Akerlof et al. 2000; Park et al. 2000; Boër et al. 2001) were designed to follow-up on GRB positions rapidly ($t \lesssim$ 1 min from GRB trigger) and proved a great success upon the discovery of a 9th magnitude optical transient following GRB 990123 (Akerlof et al. 1999).

[†] This chapter was submitted to the *Publications of the Astronomical Society of the Pacific* on 7 March 2002.

Owing to the large fields of view (typically larger than 1 deg in diameter), robotic telescopes are ideally suited to the rapid, but crude burst localizations from BATSE (5–10 deg diameter uncertainty radii). By virtue of aperture size and plate scale, the robotic telescopes are, however, limited in sensitivity and cannot provide detailed light curve information once an afterglow is older than 1–3 hours ($V \gtrsim 18$ mag). For more detailed, longer-term observations of GRBs afterglows, a readily available imager on a large-aperture telescope was clearly warranted.

Recognizing this need and potential, we built and commissioned a dual-band optical imager for the Hale 200 inch Telescope at Palomar Observatory. The instrument, named the *Jacobs Camera* (JCAM) after the private donor who provided the funds to build the instrument, has a small field–of–view (1.6 arcmin radius) and was built for under \$65 k. It is now permanently mounted (until replaced by some future instrument) at the East Arm f/16 focus and was dedicated as a rapid ($t \lesssim 30$ min from trigger) follow-up photometer of gamma-ray burst (GRB) afterglows. The advantages of JCAM were to be to uniquely provide accurate simultaneous color information while sampling the light curve on short (seconds to minutes) to long (hours to days) timescales. The instrument was designed to complement accurate ($\lesssim 3$ arcmin diameter) and rapid locations of GRBs by *HETE-II* (Ricker & HETE Science Team 2001) and later, *Swift* (Gehrels 2000).

SECTION 10.2
Scientific Motivation

Beginning with the first optical detection of a GRB afterglow in 1997 (van Paradijs et al. 1997), our group at Caltech has been involved in a campaign to locate and photometrically monitor optical afterglows associated with GRBs. This optical afterglow radiation is thought to result from synchrotron radiation generated by a relativistic blastwave as it interacts with the surrounding medium (e.g., Piran 1999). Because of the effort required to alert the community and localize optical transients, our knowledge of the early-time behavior of GRB afterglows is still sparse (cf. Akerlof et al. 1999).

Theoretical models of the afterglow emission are principally constrained by well-characterized, long time series of data taken over days to months where the temporal and spectral evolution of the afterglow is relatively mild. However, the same models predict very strong evolution in the total flux and spectral slope of the afterglow in the first 1–2 hours after the explosion when emission from the reverse shock decays and emission from the forward shock brightens.

If the synchrotron hypothesis for the origin of GRB afterglows is correct, then the early-time behavior of the afterglows should show two as yet unobserved transitions on timescales of tens of minutes to hours. Figure 10.1 depicts schematically the temporal and spectral prediction of the reverse-forward shock transition. Observations of the transitions carry an important diagnostic of the initial parameters of the GRB itself since the reverse shock liberates kinetic energy before the blastwave begins to decelerate self-similarly. Prompt observations are the only way to directly measure the initial Lorentz factor, Γ_0 (Sari & Piran 1999b). The timescales of these transitions have yet to be explored observationally.

After Akerlof et al. (1999) discovered the bright prompt optical emission of GRB 990123 attributed to a reverse shock, Kulkarni et al. (1999b) also observed a bright radio flare about one day after the burst that authors attributed to the same reverse shock. Though no other reverse shocks have been observed optically (e.g., Akerlof et al. 2000), it is now believed that at least 25% of radio afterglows show evidence for the presence of a reverse shock. Optical reverse shock signatures should be just as pervasive if observed to fainter levels than the robotic telescopes have allowed.

The degree of inhomogeneity surrounding the burst will also leave an imprint on the afterglow in the form of small deviations from power-law behavior. Wang & Loeb (2000) have predicted the degree and character of temporal variability induced by inhomogeneities in the immediate environment of the GRB. Specifically, they calculate that the r.m.s. variability at optical wavelengths should be observed at the 0.1 – 10% level about 1 hour after the GRB. The fluctuation amplitude and timescale are directly related to the length scale of the density perturbations and is essentially undetectable after a few days.

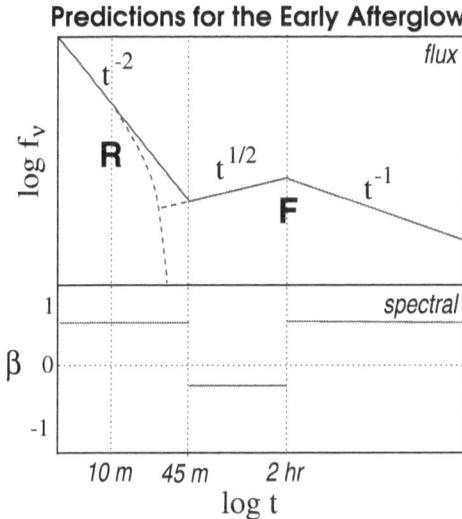

Figure 10.1 Theoretical evolution of the reverse-forward shock transition in the early afterglow. The dramatic change in spectral slope β ($f_\nu \propto \nu^{-\beta}$) during the onset of the forward shock as well as the temporal decays should be detectable by JCAM. Here "R" refers to emission from the reverse shock and "F" refers to emission dominated by the forward shock. After the reverse shock sweeps through the ejecta completely, the optical flux density will decay roughly as t^{-2}. A transition then occurs whereby emission from the forward shock dominates and the classical GRB afterglow begins. On a 45 minute timescale after the burst, the optical flux begins to be dominated by the forward shock emission and will appear blue ($\beta = -1/3$). At this time, the afterglow is expected to be in the $R \sim 15 - 18$ mag range. Later the afterglow declines as $\sim t^{-1}$ with a red spectrum ($\beta \approx 0.7$). See Sari & Piran (1999b) for details. These time dependencies and spectral evolution are distinct signatures of the synchrotron model.

Section 10.3
Instrumentation

We chose to use the East Arm f/16 focus of the Hale 200 inch since the port was available and we could be insured at least several years of permanent mounting, a timescale to coincide with the lifetime of the *HETE-II* satellite and the beginning of the *Swift* mission. Given the science goals we then formulated several practical objectives and limitations for the instrument. We required JCAM:

- To be capable of background-limited imaging in $UBVRI$ bands, with unfiltered throughput $\geq$ 65% in B, V, and R, and $> 40\%$ in U and I, inclusive of detector Q.E., exclusive of telescope throughput (that is, a 4-mirror system).

- To have a large enough image field to encompass at least one comparison star $\leq$ 17th mag in the frame with the GRB afterglow source at any Galactic latitude.

- To have image quality equivalent to r.m.s. spot diameters of $< 0''.5$ at field corners.

- To be capable of integration times from 1 s to 30 minutes with rapid response time and low latency.

- To cost < \$65,000 in total using mostly "off-the-shelf" parts.

- To be operated remotely (and efficiently) over the Internet with minimal on-site assistance.

Here we describe the construction of JCAM and discuss the rationale behind the design choices in the context of the science and practical limitations described in the previous sections. The instrument, shown in figure 10.2, was first mounted in November 2000 and finished full commissioning in February 2002.

As JCAM is now permanently mounted on the 200 inch Telescope, the observations such as those described in §10.2 are now feasible even in lunar conditions when optical imagers are not normally mounted on the telescopes of major observatories (i.e., bright time). Practically, this implies that optical imaging of the observations any afterglow localized northward of $-35°$ declination can be followed up to depths comparable with other optical imagers available less than half of the nights.

10.3.1 Optical design

A detailed listing of the optical elements of the 200 inch Telescope and JCAM are given in table 10.1. The total un-vignetted field size available at the f/16 focus of the East Arm is about 4 arcmin diameter, large enough to cover some prompt *HETE-II* (and all *Swift*) localizations and adequate to find suitable photometric and astrometric tie stars for imaging of known (i.e., precisely localized) transients.

Obtaining an acceptable field size of the instrument was a trade-off between CCD format and the amount of focal reduction. Based on our cost analysis, we opted for a rather large reduction ratio of 6:1 rather than the much more expensive large format CCD camera. This provides a near optimum scale (with respect to the typical seeing at Palomar of 1–1.5 arcsec FWHM) of 15.5 arcsec mm^{-1}, or $\sim 0''.37$ pixel^{-1} for 24μm pixels. To reduce the size of the optics which follow, the initial element is a field lens, followed by a collimator which both reduces the power required in the camera optics, and provides a collimated beam for the dichroic and filters; the field lens and the collimator have the same focal length (500 mm) and are set apart at that distance. These lenses are coated with a MgF coating, offering the best transmission performance over the broad range of wavelengths. The cameras are equipped with

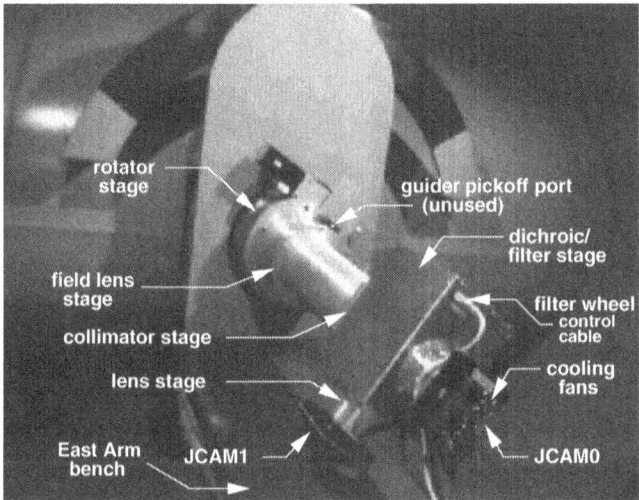

Figure 10.2 Picture of JCAM mounted at the East Arm with labeling of JCAM components; looking north-easterly from the stairs inside the East Arm. The f/16 image plane rests at the field lens stage. The light is collimated at the collimator stage, then split by the dichroic and passed through the filter wheel for each camera (JCAM0 = red side; JCAM 1 = blue side). The black and light gray cables from the two Apogee cameras and the filter wheels are connected to the JCAM computer and peripherals (not shown) lower down on the East Arm bench (see fig. 10.4).

10.3. INSTRUMENTATION

Table 10.1. Properties of Optical Elements in JCAM

Surface	Radius of Curvature (mm)	Thickness (mm)	Material	Diameter (mm)	Manufacturer	Comments
Hale 200 inch Telescope Elements						
primary mirror	33926.78	-13713.46[d]	Pyrex: Al coated	10210.8	Corning Glass	paraboloid ($k = -1$)
secondary mirror	8089.39	-8379.46[d]	Al coated	1041.4		convex hyperboloid ($k = -1.53$)
tertiary flat ("Coudé" mirror)	∞	5511.673	Al coated	[e]		
quaternary mirror	∞	1266.952	Al coated	292.1		
JCAM Elements						
field lens	1000.0	6.4	fused silica, MgF coating	101.6	JML Optical	plano-convex; FPX12080/100
collimator	1000.0	7.7	fused silica, MgF coating	50.8	JML Optical	FPX12060/100
Dichroic	∞	3.25	Float glass substrate [b]		Custom Scientific	45 deg angle of incidence[c]
Filters						
Bessel U	∞	5.0	UG, S-8612, WG305	50.0	Omega Optical	[a]
Bessel B	∞	5.0	GG385, S-8612, BG12	50.0	Omega Optical	
Bessel V	∞	5.0	GG495, S-8612	50.0	Omega Optical	
Bessel R	∞	5.0	OG570, KG3	50.0	Omega Optical	
Bessel I	∞	5.0	WG305, RG9	50.0	Omega Optical	
Sloan g'	∞	5.0	OG400, S-8612	50.0	Custom Scientific	
Sloan r'	∞	5.0	OG550	50.0	Custom Scientific	
Re-imaging Lens ($\times 2$)	170.0	proprietary	proprietary	80.5	Nikon (Nikkor)	F/1.4; 7 element lens in 5 groups

Note. — [a] All filters are specified with a maximum transmitted wavefront distortion $\lambda/4$ per inch, minimum flatness of $\lambda/4$ (or better), and are coated with dielectic anti-reflective coatings on both sides to increase transmission and reduce the effects of ghosting. [b] 4 in $\times$ 3 in rectangular. [c] Typical flatness of $4\times\lambda$ per in, parallelism of 8 arcmin, and anti-reflectance coated. [d] Negative values are for mirrors. The absolute value of this number is the distance to the next optical element. [e] 53 in $\times$ 36 in elliptical mirror.

f/1.4 85mm Nikon Nikkor lenses which likely exceed a standard telephoto lens in image quality; however, one unknown in the design process was the throughput of the lenses at ultraviolet wavelengths.

We recognized that a 6:1 focal reduction would be difficult to manage over the entire specified wavelength range with a single set of camera optics and coatings. Given this, and the science objectives, we opted to separate the blue and red light with a dichroic. After the light is split by the dichroic, each beam passes through its own filter wheel, camera lens, and CCD. However, since the field-lens and collimator are singlets, there is a chromatic dependence on the optimum focus for each filter. We found, however, that with an appropriate internal focus setting on each camera such that the Sloan r' and Sloan g' are parfocal (i.e., optimally small point-spread function r.m.s. and good image quality in both filters for a given position of the secondary mirror), as are Bessel U and Bessel I. The best setting of the focus value for secondary mirror of the telescope is consistently 0.35–0.40 mm larger for Bessel U/I than for Sloan r'/g'.

Our design target for image quality of no greater than an r.m.s. of 0.5 arcsec on the field edges appears to have been accomplished in practice as we have not measured any substantial image degradation across on-sky images; this is not unexpected given we have only had ~5 hr of imaging below 1 arcsec seeing.

Filters and Dichroic

The motorized filter wheels (Oriel part #77384), one for each camera, can carry up to five, 2 in diameter filters. The movement of the filters is controlled via TTL pulses generated from the rotator control system which, in turn, is initiated through a serial-port interface with the host computer (see §10.3.3). We purchased the pre-fabricated Bessel filters (U, B, V, R, I) from Omega Optical, Inc.[1] based in Brattleboro, VT. In addition, given the higher throughput of Sloan filters, we also decided to purchase custom designed Sloan g' and Sloan r' from Custom Scientific, Inc.[2] based in Phoenix, Arizona. The U, B, V and Sloan g' filters are installed on the blue-side camera and the remainder are installed on the red-side camera. A summary of the relevant physical specifications for the filters is provided in table 10.1.

Recently (December 2001) we replaced our initial dichroic, which was slightly undersized but purchased off-the-shelf for only ~$300, with a more expensive ($1600), higher-throughput custom built dichroic from Custom Scientific. The dichroic was fabricated and designed to accommodate the full 2 in diameter incoming collimated beam. At a 45 degree tilt angle, the dichroic, reflective at blue wavelengths and transmissive at red wavelengths, presents a 2.82 in × 3 in (height) to the incident beam. The central cut-off wavelength, nominally 5577 Å at half-power, was designed to coincide with the half-power cut-off (and turn-on) wavelengths of the Sloan g' and Sloan r' filters. Given the large size of the dichroic, and the fact that the optical bread-board can be turned upside-down at certain rotator angles, we designed and built a custom holder to secure the dichroic.

The effective filter transmission curves are shown in figure 10.3, created by the convolution of the quantum efficiency curve of the CCD and the transmission of the filters and dichroic. We provide the tabular version of these curves as well as estimates of the transmission curves through the entire telescope + JCAM system in table 10.6. In table 10.2 we provide a summary of the basic wavelength properties of the curves, such as FWHM and peak response efficiency, as well as a spectrum-dependent tabulation of the effective wavelength of the filters as a function of airmass. Table 10.3 shows the synthetic JCAM AB magnitudes of two primary stars in the Sloan filter system tabulated using equation 7 of Fukugita et al. (1996).

Detectors

Both detectors are Apogee Ap-7b, with a SITe thinned, backside-illuminated 512x512 CCD with 24 micron pixels. The CCDs are thermoelectrically cooled and, depending on the ambient temperature

---
[1] http://www.omegafilters.com/
[2] http://www.CustomScientific.com/

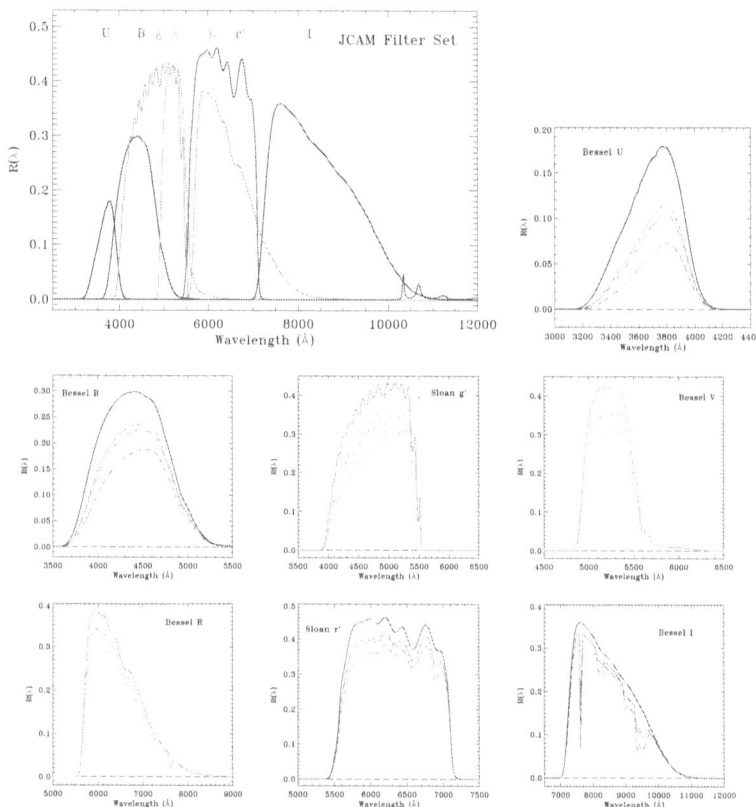

Figure 10.3 The calculated filter response curves of JCAM. Shown at top left are the effective response curves through the 7 filters of JCAM (U, B, g', V, R, r', I) found using the laboratory transmission curves of the filters themselves, the laboratory transmission and reflection curves of the dichroic, the nominal quantum efficiency of the SITe backside illuminated CCD, and the nominal reflectivity of aluminum (4 mirror elements). The effect of the remainder of the optical elements—field lens, collimator, 7-element Nikon lens (all of which are anti-reflective coded)—are each modeled as a 98.5% transmissive element. The individual response curves are also shown with the solid curve representing the instrument response alone. The other curves in each plot correspond to the calculated response at an airmass of 1.0, 1.2, and 2.0 (dash-dot, dash-dot-dot-dot, and dashed, respectively) using an average atmospheric transmission curve for Palomar. The small "red leak" at ~ 1.05 micron is from the Sloan r' filter.

Table 10.2. Summary of JCAM Filter Properties

Filter	Side[a]	λ_{peak}[b] Å	$R(\lambda_{peak})$[b]	Airmass	FWHM Å	λ_{eff} (Å)[c] $\beta=-3.0$	-2.5	-2.0	-1.5	-1.0	-0.5	0.0
Bessel U	1	3787.5	0.179	0.0	453.2	3712.3	3705.9	3700.3	3695.1	3690.2	3685.4	3680.7
				1.2	422.3	3735.5	3728.3	3722.2	3716.8	3711.8	3707.1	3702.5
				2.0	406.1	3751.0	3742.9	3736.2	3730.5	3725.3	3720.6	3716.1
Bessel B	1	4411.5	0.299	0.0	945.1	4428.7	4416.1	4403.6	4391.3	4379.1	4366.9	4354.8
				1.2	919.5	4455.6	4443.2	4431.0	4418.9	4406.8	4394.8	4382.9
				2.0	899.0	4472.8	4460.6	4448.5	4436.6	4424.7	4412.8	4401.0
Sloan g'	1	5095.0	0.432	0.0	1320.7	4827.8	4811.9	4795.7	4779.4	4762.9	4746.4	4729.8
				1.2	1268.8	4849.8	4834.3	4818.7	4802.9	4786.9	4770.8	4754.6
				2.0	1226.7	4863.7	4848.6	4833.4	4817.9	4802.2	4786.4	4770.5
Bessel V	1	5171.0	0.423	0.0	560.7	5261.2	5256.8	5252.5	5248.3	5244.3	5240.3	5236.3
				1.2	562.0	5263.8	5259.4	5255.0	5250.8	5246.7	5242.7	5238.7
				2.0	562.4	5265.6	5261.1	5256.7	5252.5	5248.3	5244.3	5240.3
Bessel R	0	5929.0	0.386	0.0	1209.7	6490.8	6467.1	6444.1	6421.8	6400.1	6379.2	6359.0
				1.2	1228.2	6498.2	6474.6	6451.8	6429.6	6408.1	6387.3	6367.2
				2.0	1239.5	6503.9	6480.4	6457.6	6435.5	6414.1	6393.3	6373.2
Sloan r'	0	6197.5	0.462	0.0	1498.8	6420.8	6387.5	6358.1	6331.9	6308.2	6286.5	6266.4
				1.2	1496.1	6436.2	6401.8	6371.6	6344.8	6320.6	6298.5	6278.1
				2.0	1493.7	6446.5	6411.4	6380.7	6353.4	6328.9	6306.5	6285.9
Bessel I	0	7617.0	0.359	0.0	2085.0	8506.1	8467.4	8429.4	8392.0	8355.4	8319.7	8284.8
				1.2	1886.2	8486.1	8447.8	8410.3	8373.6	8337.8	8302.9	8268.8
				2.0	1792.2	8475.1	8437.0	8399.8	8363.5	8328.0	8293.5	8259.9

[a]The index corresponds to the camera number for that given filter. An index of "0" corresponds to the red side camera (transmissive through the dichroic) and an index of "1" corresponds to the blue side camera.

[b]The peak efficiency of the response curve $R(\lambda)$ (col. 4) and the corresponding wavelength (col. 3) computed before including a wavelength-dependent atmospheric effect of the filter curves (i.e., airmass equal zero). The efficiency curves are tabulated using laboratory transmission measurres of the individual filters and the dichroic. The quantum efficiency of the CCDs are taken from manufacturer data sheet for the SITe 512x512 24μm chip. In addition, to account for the 4 mirrors in the system, we use the wavelength-dependent nominal reflectivity of aluminum as presented on page 12-117 of Lide (1994). We assume a wavelength independent transmission (T = 98.5%) for the remaining 9 optical elements (field lens + collimator + 7-element lens). On-sky imaging suggests that the peak response (col. 4) should be scaled downward by 3.170, 1.62, 1.53, 1.38, 1.75, 1.71, 3.54 in the filters U, B, V, g', R, r', and I, respectively. See text for an explanation.

[c]The effective wavelength (λ_{eff}) for a given filter at the given airmass and input object spectrum. An average wavelength-dependent atmospheric transmission curve at Palomar is assumed. A power-law input object spectrum is assumed, with $f_\nu \propto \nu^{-\beta}$. The usual definition of λ_{eff} (see Fukugita et al. 1995, equation A1) is for airmass = 0.0 and $\beta = -2$ (i.e., source flux f_λ = constant).

Table 10.3. Synthetic Magnitudes of Primary Standard Stars Through the JCAM Filter Set

Filter	BD+26°2606			α Lyr		
	$m_{\text{AB, JCAM}}$ mag	λ_{eff} Å	Δm mag	$m_{\text{AB, JCAM}}$ mag	λ_{eff} Å	Δm mag
Bessel U...	10.557	3735.812		0.596	3770.95	
Bessel B...	10.030	4416.799		-0.119	4382.95	0.009
Sloan g'...	9.892	4796.175	0.002	-0.092	4728.26	-0.005
Bessel V...	9.750	5246.947		-0.025	5232.20	
Bessel R...	9.584	6383.269		0.192	6324.35	0.007
Sloan r'...	9.592	6302.615	0.018	0.176	6242.14	0.013
Bessel I...	9.504	8279.067		0.455	8233.87	0.001

Note. — The synthetic AB magnitudes of the primary standards of the Sloan system are calculated using the observed flux of the stars (as given in table 6 of Fukugita et al. 1996) convolved with the calculated instrumental response. The effective wavelength λ_{eff} of the filter (calculated following equation 3 of Fukugita et al. 1996) is also provided. The offset of the synthetic values for some filters (cols. 4 and 7) are found using the magnitudes presented in Fukugita et al. (1996) and Fukugita et al. (1995). The good agreement suggests that the JCAM Sloan filters should very well approximate the same SDSS filters. The absence of a detectable color term using on-sky data corroborates this statement (see §10.4.3).

inside the dome of the 200 inch, can typically reach temperatures of -40 deg C after cooling for 15 minutes. Since JCAM is permanently mounted, the self-contained cooling allows for a relatively maintenance-free system which can, in turn, be brought to a ready state by a remote observer (that is, without the need for on-site filling of cryogenic dewars). The total cost for both CCD systems was $16 k (as compared with $\gtrsim$ $200 k for custom systems), greatly reducing the overall cost of JCAM.

The temperature and shutter control, as well as CCD binning modes, are controlled through an ISA bus card which connects to the CCD housing through a shielded cable. The A/D converters reside on the CCD housing and the ISA card provides a buffer for the data before they are read to disk. A full-chip readout, including 30 extra bias ("overscan") lines, requires 9.5 sec per detector. Though the detectors may be readout "simultaneously" (13 sec total) such a process introduces an apparent cross-talk which results in erratic and high levels of read-noise on certain columns. We were unable to isolate and remove the source of the cross-talk but it likely occurs in the interaction on the ISA bus since the effect can be mimicked by forcing CPU interrupts during the readout.

The measured read noise and gain are given in table 10.5 and are in line with the manufacturer's specifications. In table 10.5 we give some relevant characteristics of the JCAM CCDs and images. Note that the gain in both CCDs provide an adequate Nyquist sampling of the read noise. The linearity of both CCDs is acceptable to about 55 k DN.

Table 10.4. Summary of GRB Triggers Observed to Date with JCAM

GRB/Trigger Name	Δt	Conditions	Result/ Comments	References
GRB 020124	1.88 day	thick cirrus	detection at $r' \approx 24$ mag	Bloom (2002)
GRB 011211	6.73 hr	very cloudy, $4''$ seeing	non-detection of candidate solidifies transient	Bloom & Berger (2001)
HETE # 1793	1.62 hr	clear	false trigger from solar flare; not a GRB	
GRB 010222	8.16 day	light cirrus	detection of OT in 4 bands $R = 23.19 \pm 0.15$ mag	fig. 10.6; Bloom et al. (2002)
GRB 001018	60.6 day	thick cirrus, $2.8''$ seeing	non-detection of host to $R = 22.7$ mag	Bloom et al. (2001)

10.3.2 Mechanical design and construction

The design of the mechanical systems of JCAM dealt with three major constraints. First, the optics and physical dimensions of the elements required the field lens to be ∼5 inches from the entrance to the East Arm port and the CCDs to be at least ∼30 inch from the field lens. Second, the entire system needed to rotate so as to observe at a cardinal position angle[3] yet there is only ∼13 inch clearance from the center of the optical axis to the East Arm bench (see fig. 10.2). Third, the East Arm bench and the optical axis are aligned toward the pole and so the entire system must operate at 33 deg angle from level without significant flexure.

The bench clearance constraint and optical distances implied that the two CCDs would have to be back-loaded at the end of the optical axis if the instrument were to rigidly rotate. This implied that the rotator stage would have to carry a heavy load yet be capable of fine angular positioning. We purchased a stepper-motor rotator stage from Newport which offers a large load capacity, an 11" central clearance, and angular positioning resolution of 0.001 degrees. This stage is mounted just before the field lens stage (fig. 10.2).

The remainder of the stages were custom designed by us and constructed out of aluminum at machine shops at the University of North Carolina and Caltech. The data-taking computer (see §10.3.3), rotator control module and filter-wheel staging electronics are all housed in a rack which we mounted to the East Arm bench just below JCAM.

10.3.3 Electronics and software implementation

In the interest of simplicity, we decided that all camera, rotator, and filter operations would be conducted through a single computer which would be remotely controlled over the Internet. A schematic of the basic software and electronics configuration is shown in figure 10.4. The computer is currently running Linux RedHat version 7.2 (kernel 2.4.2) and is accessed via a secure (SSH2) connection. In the interest of security, we periodically update the kernel and connection software.

The data acquisition and hardware are controlled by a single Dell OptiPlex GX1 MiniTower 450 MHz Pentium III computer with 20 GB of IDE disk space. A co-axial network cable inside the East Arm connects the computer (automatically upon boot) to the Palomar network at data transfer rates of ∼2 Mbps. The specific model of the computer was chosen since it allows up to 4 ISA card slots; JCAM uses only 2 of these slots for the Apogee control cards but we wanted to keep the machinery upgradeable. The power for the two Apogee cameras, supplied though the ISA cards, is provided from the computer; despite the current load from the cameras (1.4 amp per camera) the computer power

---
[3] We originally required the field rotation so that we could be assured that the pick-off guide camera could observe a bright enough guide star somewhere in the circular annulus about the science field center.

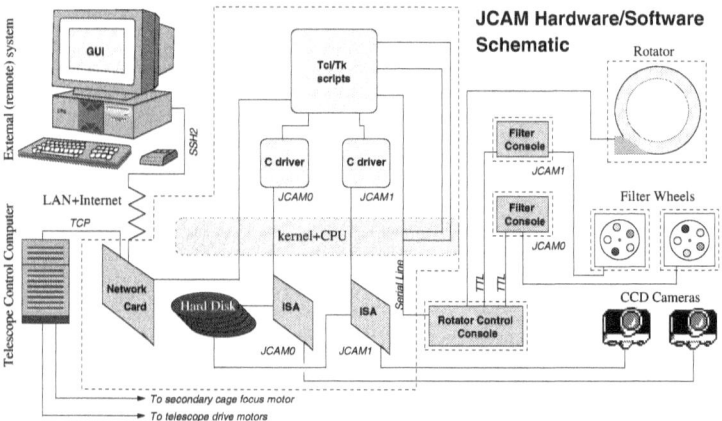

Figure 10.4 Schematic of the hardware and software configuration of JCAM. The main computer, located inside the East Arm rack, is responsible for the interaction with the Apogee CCD cameras, the control of the rotator and filter wheels, data taking and storage, and communication with the user via a secure SSH connection over the Internet. The interaction of the user (via a GUI on the remote machine; see fig. 10.5) with the instrument operations is controlled though Tcl/Tk and C codes (see text) and routed through the Linux kernel/CPU (depicted here as a filled gray block). Separate physical entities (e.g., the JCAM computer, filter wheels) are encapsulated here in dashed lines.

supply performs well and there is no apparent pattern noise on the CCD dark frames.

The interaction with the rotator and filter wheels is conducted through a serial port interface with the rotator control stage from Newport. The control software sends short ASCII string commands to this stage and can poll the stage for the status of the rotator. The control software also initiates the generation of TTL pulses from the stage used to interact with the two filter wheels.

The low-level CCD control software—the device driver interface between the kernel and the ISA cards—was written in C and purchased as source code from the ClearSkyInstitute. With each new kernel update, we have modified this code to comply with the protocol for the way that device drivers interact with the kernel. In the future, we may begin to use a new driver source, called Linux Apogee Instruments camera drivers[4] For the control of the device drivers, installed at boot as two independent kernel modules (one for each camera), we have written C-code wrappers and a set of Tcl/Tk (current version 8.3) scripts. This software, amounting to about 10 k lines of code, is also used to interact with the telescope control system (TCS) via a TCP connection, control the filter wheels and rotator, and to write data acquired from the CCDs to disk. The user interacts with the code through a graphical user interface (GUI) (also written in Tcl/Tk) which is displayed on the remote (external) computer monitor though the SSH port.

SECTION 10.4
Operations and Performance

JCAM is designed to operate in a "Target of Opportunity" (ToO) mode, temporarily interrupting the scheduled observer if a GRB afterglow or other transient event is to be observed rapidly. Some

---

[4] See http://www.randomfactory.com/apogee-lfa.html

reconfiguration of the telescope may thus be necessary before on can start taking data using JCAM.

10.4.1 Initiating operations

Here we outline the steps to begin observing with JCAM and the approximate time required per step. Notification of burst alerts reaches our group members via cellular phone and email messaging through the GRB Coordinates Network (GCN[5]). *Swift* alerts will be nearly simultaneous with burst trigger, but the time delay for *HETE-II* alerts depends on the precise location of the satellite with respect to ground stations.[6] Typically we receive notice of an approximate position less than one minute after the burst trigger and can then initiate a ToO by way of a phone call to the observers within the next few minutes.

Running in ToO mode, the Coudé mirror may not yet be deployed and so must to be swung into place by the telescope operator before JCAM operations can commence. To do so, the telescope first is stowed at zenith; the slew to zenith can take between 1–4 minutes depending upon the initial pointing. The telescope operator then operates the Coudé crane from a console on the catwalk floor. The Coudé crane places the tertiary mirror atop the Cassegrain baffling tube and then the tertiary is rotated by 90 deg about the Cassegrain tube so as to point at the East Arm entrance hole. This process requires about 22 minutes. The secondary mirror is then swung into place, if it is not in place already, and set to a nominal secondary focus distance 58mm. This process requires about 4 minutes.

During the mechanical reconfiguration, the remote observer logs into the JCAM computer and starts the JCAM software which then begins to cool the CCDs and establish the connections to the TCS, rotator, and filter wheels. This process takes a total of ∼5 minutes with an additional 10–20 minutes for the CCDs to cool fully. Once the mirrors are in place, the telescope is slewed to a bright SAO star near the science target to check pointing. We then perform a focus loop on the SAO star requiring about 5–10 minutes. These operations can be performed before the CCDs are fully cooled. In total, when the entire telescope must be reconfigured, as in this example, the overhead time from start to science target requires about 45 minutes.

In the case where the Coudé mirror has been put in place during the daytime (i.e., when the observers are using only Prime focus as the primary science instrument) the overhead "time to science target" can be reduced substantially to ∼15 minutes. Here only the zenith slew and secondary mirror flip, plus on-sky set up are required.

Though some science observations can likely commence within 15 minutes increasing the likelihood that we will observe the transition from reverse to forward shock (fig. 10.1), the 45 minute timescale for longer ToO turn-around is comparable to this timescale. We note, however, the transition time is a sensitive function of the initial Lorentz factor of the explosion and the redshift of the GRB so it may still be possible to observe this transition. In either scenario, the observations of the peak of the forward shock (fig. 10.1) and the small variability will be possible.

10.4.2 Data taking and observing procedures

All data and instruments are controlled with a simple front-end GUI. We have also written a hand-paddle and secondary focus tool to facilitate dithering and focus changes that minimize the need to communicate with the telescope operator. A screenshot example of the GUI is shown in figure 10.5.

All of the imaging data are stored on a data partition of the JCAM computer hard drive. Usually, after the data are written to disk, the remote observer copies the data to the remote machine to inspect the images; this can be done individually or by periodically synchronizing the remote and JCAM data directories (using, for example, the password-encrypted RSYNC command). As the images are rather small (527 MB uncompressed; 150–400 MB compressed) the transfer times require between 1–25 s per

---
[5] See http://gcn.gsfc.nasa.gov/gcn/
[6] See http://space.mit.edu/HETE/Bursts/

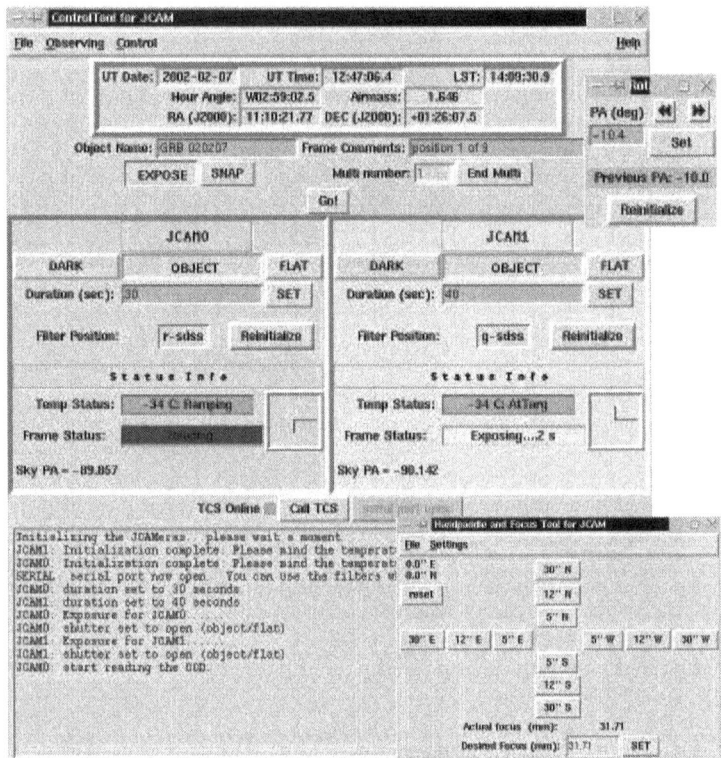

Figure 10.5 Screenshot of the JCAM graphical user interface (GUI). In the main window of the GUI, a real-time summary of the current pointing, time, etc., is provided at top. Object name and frame comments are entered by the user and the observing mode ("EXPOSE" for beginning exposures on both CCDs or "SNAP" to begin the exposures separately). The filter, integration time, and frame type are chosen individually for each camera. A status display for each camera shows the CCD temperature, exposure status, and current position angle of the exposure. A small compass rose shows the sky orientation of the CCD, with the longer line pointing North and the shorter line indicating East. Due to the dichroic, there is a y-axis flip between the two CCDs. At the bottom of the main window is a text log, which is also saved with time-stamps to a log file. Offset from the main window are two smaller windows, one to control the rotator angle (top) and the other to aid in small changes in the telescope positioning (dithering) and secondary focus (bottom).

image, limited only by the bandwidth between Palomar and the remote observer site. The typical transfer time between Palomar and Pasadena is 4 s for an uncompressed image.

We decided not to install the guider camera after realizing, by comparison with the point-spread function of images obtained contemporaneously in the same filter at the 60 inch telescope, that 200 inch Telescope tracks very well over times $\lesssim$ 300 s. Aside from an occasional smearing due to telescope jump upon wind-shake (which equally degrades guided-images) the image quality on long exposures is comparable to the image quality of short exposures. Nevertheless, to minimize tracking errors, we typically integrate for shorter time periods than is usual with larger-format guided imaging.

The shorter exposure times, typically integration 100–250s, is warranted by the short read times (9.5 s) and the fact that the frames are sky-limited after $\lesssim$2 sec of integration time, except in U-band where the sky dominates after about 30 s; the exact times depend, of course, on the sky brightness contribution from the moon. More frames per field, when dithered between exposures, also allow for the construction of better supersky flat fields and the removal of cosmic rays and CCD defects. When the seeing and/or transmission changes rapidly, more images can also be useful in constructing a higher signal-to-noise summed image of the field (that is, by giving lower weight to those frames with lower signal-to-noise detection of point sources). Note that thanks to the large collecting area of the 200 inch telescope, *JCAM imaging is never dark current limited* despite being a thermoelectrically cooled system.

Depending on the science objectives, we typically observe at least 5 frames simultaneously in the Sloan r' and g' for 100 s and 110 s, respectively. The 10 s difference in exposure time allows JCAM1 (Sloan g') to finish exposing just as JCAM0 (Sloan r') finishes readout. For the other par focal set, Bessel I and Bessel U, we typically acquired two 100 s frames in I band while exposing U band for 200 s. (Note that the filters B/V and R are largely superseded in efficiency by the filters Sloan g' and r', respectively; see fig. 10.3). We have two modes of taking many frames automatically ("Multi" in fig. 10.5), one which begins an exposure as soon as the camera is finished reading out and the other which opens the shutters of both cameras simultaneously. The latter mode, when the exposure times of both cameras are equal, is particularly useful when conditions are non-photometric since both images are exposed through the same cloud pattern, preserving the relative flux of objects in the two bandpasses.

As a result of shorter exposure times, a given field requires more exposures for a given depth. In a typical full night of science imaging, we have typically generated between 600–800 images. We found that the rate of frame acquisition is too high to adequately log the frames by hand. As such, we wrote an electronic logbook program in Tcl/Tk which automatically creates a new line for each new image that is acquired. The pertinent header information is shown and the user can add comments to each line and then save the logbook to text, Postscript, and graphical output. Aided by the existence of electronic logs, we have begun to archive each observing run in a uniform set of web pages.[7]

10.4.3 Preliminary results

Despite poor observing conditions on all but one night of commissioning over the past year (10 nights), we managed to observe a number of GRB and GRB-related targets during the commissioning period of JCAM. The results of some of these observations are summarized in table 10.4; a JCAM dual-band image of GRB 010222 is shown in fig. 10.6. On two occasions, observers have successfully imaged a GRB position by operating JCAM from a remote location, GRB 011211 (from Waimea, HI) and GRB 020124 (from Pasadena, CA).

As can be seen from the table, we have unfortunately not yet observed a *bona fide* GRB position on rapid turn-around timescale ($\lesssim$ 1 hr). The only source which we followed-up rapidly (HETE #1793) turned out to be a particle event in the *HETE-II* detectors generated after solar flare activity. The disappointing absence of rapid follow-up events is due to the low rate of GRBs with rapid positional determination and the absence of localizations to accuracies smaller than $\sim$ 5 arcmin diameter.

---

[7] http://www.astro.caltech.edu/~jsb/Jcam/runsum/

Table 10.5. JCAM Detector Characteristics

Pixel size...	24 μm (0″.371)
CCD size...	512 × 512 pix^2
including overscan regions	535 × 533 pix^2
Field of View...	3′.17 × 3′.17
Avg. Read time...	9.5 s: single-frame mode
	14 s: dual-frame mode
Inverse Gain...	4.2 electron DN^{-1} (JCAM0)
	3.9 electron DN^{-1} (JCAM1)
Read noise (r.m.s.) ...	8.5 electron (JCAM0)
	9.0 electron (JCAM1)
Dark current ...	0.16 electron pix^{-1} s^{-1} (JCAM0)
	0.18 electron pix^{-1} s^{-1} (JCAM1)

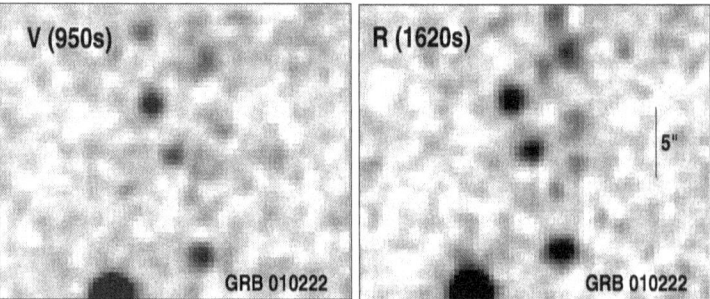

Figure 10.6 JCAM images of the afterglow of GRB 010222 (object at center). The transient was faint ($V = 23.54$ mag, $R = 23.19$ mag) at the time of observation (8 days after the GRB) but was detected at the 20-σ level in roughly 15 minutes of integration through light cirrus.

Given the rapid readout and its ability to perform dual-band simultaneous photometry, the instrument has also been in demand for projects requiring deep color snapshots at a high duty-cycle (e.g., Kuiper Belt Object surveys, distant galaxy cluster surveys). The instrument would also be well-suited for photometric monitoring programs of faint variable sources (e.g., gravitational lenses) which require short pointed observations over a large number of nights. Indeed, by reaching limiting magnitudes of 23–25 every 10 minutes, a typical night with JCAM could be used to image ∼30 faint sources in two colors with the same astrometric and photometric quality as other larger-format optical imaging instruments.

Observed System Throughput

On 4 Feb 2002, we observed several secondary standard stars through a variety of airmasses in photometric conditions. The stars, spanning a large range in colors, were selected from the calibrated Sloan catalog of standards (Smith et al. 2002) as those near to the science targets of the night. For all seven filters, we fit for the zeropoint, color term(s), and airmass extinction to the SDSS filter system. The transformations can be summarized as follows:

$$\begin{aligned}
u' &= (22.435 \pm 0.023) - (0.765 \pm 0.052) \times (X-1) \\
&\quad -(1.662 \pm 0.019) \times (u'-g'-1) \\
&\quad -2.5\log_{10}(R_{U,\mathrm{JCAM}}),
\end{aligned}$$

$$\begin{aligned}
g' &= (25.976 \pm 0.014) - (0.350 \pm 0.011) \times (X-1) \\
&\quad -(0.206 \pm 0.020) \times (g'-r'-1) \\
&\quad -(0.115 \pm 0.014) \times (u'-g'-1) \\
&\quad -2.5\log_{10}(R_{B,\mathrm{JCAM}}),
\end{aligned}$$

$$\begin{aligned}
g' &= (26.971 \pm 0.030) - (0.323 \pm 0.042) \times (X-1) \\
&\quad +(0.013 \pm 0.029) \times (g'-r'-1) \\
&\quad -2.5\log_{10}(R_{g',\mathrm{JCAM}}),
\end{aligned}$$

$$\begin{aligned}
g' &= (26.500 \pm 0.020) - (0.387 \pm 0.065) \times (X-1) \\
&\quad +(0.370 \pm 0.008) \times (g'-r'-1) \\
&\quad -2.5\log_{10}(R_{V,\mathrm{JCAM}}),
\end{aligned}$$

$$\begin{aligned}
r' &= (26.903 \pm 0.004) - (0.181 \pm 0.007) \times (X-1) \\
&\quad +(0.002 \pm 0.004) \times (g'-r'-1) \\
&\quad -2.5\log_{10}(R_{r',\mathrm{JCAM}}),
\end{aligned}$$

$$\begin{aligned}
r' &= (26.576 \pm 0.003) - (0.134 \pm 0.004) \times (X-1) \\
&\quad +(0.039 \pm 0.002) \times (g'-r'-1) \\
&\quad -2.5\log_{10}(R_{R,\mathrm{JCAM}}),
\end{aligned}$$

$$\begin{aligned}
i' &= (25.893 \pm 0.003) - (0.101 \pm 0.006) \times (X-1) \\
&\quad +(2.124 \pm 0.011) \times (r'-i') \\
&\quad -2.5\log_{10}(R_{I,\mathrm{JCAM}}),
\end{aligned}$$

with X equal to the airmass and $R_{Y,\mathrm{JCAM}}$ equal to the object flux rate as measured through the filter Y in units of electrons per second. The uncertainties do not include the systematic uncertainties in the computed aperture correction (which are less than 0.01 mag). The zeropoints are expected, of course, to vary from night to night but the color and extinction curves should remain fairly stable. As expected, as there is no statistically significant color term, the JCAM r' and g' filters very closely resemble those of the SDSS filters. This was also confirmed by the concordance of the synthetic JCAM and SDSS magnitudes of primary standards (table 10.3). The other filters do have statistically significant color terms relative to the SDSS photometric system. Note that there is no need for a "red leak" correction to the U-band magnitude (as required with Sloan), one extra benefit of using a dichroic. We do not yet have enough photometric data to directly compute the color terms for the JCAM filter set relative to other photometric systems.

A sky brightness measurement was made at an airmass of $X = 1.73$ and 77.6 deg from a half-illuminated moon. The sky flux was 2.5, 61.1, 102.7, 66.8 electrons s^{-1} pix^{-1} in the U, g', r', and I filters, respectively. This is typical of a bright moon-lit night, and represents a nominal upper limit to the expected sky brightness levels.

Using the response curves and the measured fluxes of primary standards, we compute the total instrument response as a scaling to the calculated curves. First, we sample the standard star spectrum of BD+26°2606 (Fukugita et al. 1996) and the response curves at $\Delta_\lambda = 3$ Å intervals through a spline interpolation, converting the spectrum to flux, $S_\nu(\lambda_i)$, in units erg s^{-1} cm^{-2} Hz^{-1}. Here i refers to the bin number of the sample. The total expected count rate in the Y filter is then,

$$R_{Y,\text{exp}} = A_{\text{eff}} \sum_i \frac{S_\nu(\lambda_i)\, R_Y(\lambda_i)\, \Delta_\lambda}{h\, \lambda_i}, \tag{10.1}$$

where $R_Y(\lambda_i)$ is the tabulated response curve of filter Y at an airmass of unity, $A_{\text{eff}} = 1.93 \times 10^5$ cm^{-2} is the effective collecting area of the 200 inch Telescope, and h is Planck's constant. If the calculated response curve is correct, we expect $R_{U,\text{exp}} = 8.606 \times 10^5$, $R_{B,\text{exp}} = 5.114 \times 10^6$, $R_{V,\text{exp}} = 5.269 \times 10^6$, $R_{g',\text{exp}} = 1.039 \times 10^7$, $R_{R,\text{exp}} = 1.040 \times 10^7$, $R_{r',\text{exp}} = 1.419 \times 10^7$, and $R_{I,\text{exp}} = 1.281 \times 10^7$ electrons s^{-1} from BD+26°2606. Using our observed data we actually observed $R_{U,\text{obs}} = 2.715 \times 10^5$, $R_{B,\text{obs}} = 3.153 \times 10^6$, $R_{V,\text{obs}} = 3.447 \times 10^6$, $R_{g',\text{obs}} = 7.528 \times 10^6$, $R_{R,\text{exp}} = 5.959 \times 10^6$, $R_{r',\text{obs}} = 8.281 \times 10^6$, and $R_{I,\text{obs}} = 3.620 \times 10^6$ electrons s^{-1} from BD+26°2606. This implies that our response curves are overestimated by 3.170, 1.62, 1.53, 1.38, 1.75, 1.71, 3.54 in the filters U, B, V, g', R, r', respectively. The scale factors of ~ 1.6 in the B, V, g', R, r' filters are reasonable and likely due to a lower throughput of the 9 optical JCAM elements than assumed (98.5%) and dusty telescope mirrors (these observations were taken $\sim$10 months after re-aluminization). For example, the scale factor can be reproduced if each of these 13 elements are $\sim$4% less efficient than assumed. The large scale factor in I-band is probably due to a less efficient CCD than assumed. The large scale factor in U-band is probably due to a combination of effects of poor UV response from the telescope mirrors (for example, before re-aluminization, the U-band efficiency is typically down by $\sim$20% per mirror) and a lower UV throughput of the optical elements in the Nikon lenses.

10.4.4 Deficiencies

In the construction and operation of JCAM we have realized a number of deficiencies with the system, some expected and some not. The throughput in the near-UV, owing to the Nikon lens, is not as high as we had required but we do not view this as a major impediment to the science we hope to accomplish with the instrument. Over the past year, the T1-line Internet connection to the mountain has been largely reliable but on a few occasions the network has been slow or down. Due to weight balance restrictions, the Coudé mirror cannot be put in place when the adaptive optics (AO) system is mounted at Cassegrain focus. Unfortunately, as the frequency of the AO system use has increased, the availability of JCAM for science imaging is decreasing (now about 90% available).

Of greatest concern is the vignetting in the system which reduces the effective throughput on the edges of the JCAM fields to as low as 40% from the peak throughput near the center of the field. In practice, though, since the vignetting pattern is stable, we have developed a methodology for data reduction where the pattern may be found and removed. This leads to images where the noise is dependent upon position on the chip.

10.4.5 Future extensions

We are looking into design changes to future minimize the vignetting problem. We have also begun work to make JCAM fully scriptable; that is, to be able to run JCAM from a command-line interface rather than via a GUI. We are also writing a number of other observing tools, such as a graphical representation of the dither pattern on a given field and an extension to the electronic log program which periodically grabs water-vapor and IR weather maps from the Internet and saves them to a local disk with timestamps. The Tcl/Tk GUI is currently run by the remote user on the JCAM computer, unnecessarily taxing the CPU and memory of the computer. We are, however, examining software design changes that would allow the remote user to have a local client which controls the GUI, which in turn sends and receives small command packets to the JCAM computer (acting as a server).

For now, the bandwidth for image transfer is limited by the bandwidth of the T1 line off the mountain. With future upgrades to the site Internet connection, the current co-axial Ethernet connection of the JCAM computer to the Palomar LAN will become the limiting factor. As such, we have purchased fiber optic electronics to put the JCAM computer on the Palomar LAN at optical fiber connection rates, i.e., a factor of 100 times faster than currently. We plan to install the optical fiber system once the bandwidth for the Internet connection to Palomar is increased by a factor of ~ 10.

We are indebted to the financial and intellectual generosity of M. Jacobs without whom the project would have never been made possible. The staff and directors of Palomar Observatory are applauded and thanked for their tireless effort to help the JCAM project become as successful as possible: R. Ellis, W. Sargent, R. Brucato, R. Thicksten, R. Burress, H. Petrie, J. Henning, M. Doyle, J. Mueller, D. Tennent, S. Kunsman, and J. Phinney. At Caltech, we thank T. Small and J. Yamasaki for their sagely advice at crucial times during the construction of JCAM. We especially thank D. Fox, D. Reichart, E. Berger, and J. Eisner for assistance during commission. JSB gratefully acknowledges the fellowship and financial support from the Fannie and John Hertz Foundation. SRK acknowledges support from NASA and the NSF. AD was supported by a Millikan Fellowship at Caltech.

10.4. OPERATIONS AND PERFORMANCE

Table 10.6. JCAM Filter Response Curves

$R(\lambda)$

λ (Å)	Bessel U 0.0	Bessel U 1.2	Bessel B 0.0	Bessel B 1.2	Sloan g' 0.0	Sloan g' 1.2	Bessel V 0.0	Bessel V 1.2	Bessel R 0.0	Bessel R 1.2	Sloan r' 0.0	Sloan r' 1.2	Bessel I 0.0	Bessel I 1.2
3080	⋯	⋯	⋯	⋯	⋯	⋯	⋯	⋯	⋯	⋯	⋯	⋯	⋯	⋯
3160	5.57E-4	1.26E-4	⋯	⋯	⋯	⋯	⋯	⋯	⋯	⋯	⋯	⋯	⋯	⋯
3240	8.20E-3	2.74E-3	⋯	⋯	⋯	⋯	⋯	⋯	⋯	⋯	⋯	⋯	⋯	⋯
3320	2.89E-2	1.15E-2	⋯	⋯	⋯	⋯	⋯	⋯	⋯	⋯	⋯	⋯	⋯	⋯
3400	5.75E-2	2.52E-2	⋯	⋯	⋯	⋯	⋯	⋯	⋯	⋯	⋯	⋯	⋯	⋯
3480	8.66E-2	4.10E-2	⋯	⋯	⋯	⋯	⋯	⋯	⋯	⋯	⋯	⋯	⋯	⋯
3560	1.18E-1	5.99E-2	1.01E-4	5.14E-5	1.37E-6	⋯	⋯	⋯	⋯	⋯	⋯	⋯	⋯	⋯
3640	1.47E-1	7.91E-2	3.95E-3	2.12E-3	3.08E-6	1.66E-6	⋯	⋯	⋯	⋯	⋯	⋯	⋯	⋯
3720	1.68E-1	9.45E-2	2.47E-2	1.39E-2	4.13E-6	2.33E-6	⋯	⋯	⋯	⋯	⋯	⋯	⋯	⋯
3800	1.79E-1	1.05E-1	7.20E-2	4.21E-2	6.59E-6	3.85E-6	⋯	⋯	⋯	⋯	⋯	⋯	⋯	⋯
3880	1.43E-1	8.70E-2	1.32E-1	8.06E-2	1.23E-3	7.46E-4	⋯	⋯	⋯	⋯	⋯	⋯	⋯	⋯
3960	7.30E-2	4.66E-2	1.88E-1	1.20E-1	3.92E-2	2.50E-2	⋯	⋯	⋯	⋯	⋯	⋯	⋯	⋯
4040	2.02E-2	1.34E-2	2.30E-1	1.53E-1	1.22E-1	8.14E-2	⋯	⋯	⋯	⋯	⋯	⋯	⋯	⋯
4120	2.85E-3	1.96E-3	2.60E-1	1.78E-1	2.03E-1	1.39E-1	⋯	⋯	⋯	⋯	⋯	⋯	⋯	⋯
4200	2.57E-4	1.81E-4	2.80E-1	1.97E-1	2.55E-1	1.80E-1	⋯	⋯	⋯	⋯	⋯	⋯	⋯	⋯
4280	2.00E-5	1.44E-5	2.92E-1	2.11E-1	2.94E-1	2.12E-1	⋯	⋯	⋯	⋯	⋯	⋯	⋯	⋯
4360	⋯	⋯	2.97E-1	2.19E-1	3.28E-1	2.42E-1	1.29E-6	⋯	⋯	⋯	⋯	⋯	⋯	⋯
4440	⋯	⋯	2.98E-1	2.24E-1	3.62E-1	2.72E-1	⋯	⋯	⋯	⋯	⋯	⋯	⋯	⋯
4520	⋯	⋯	2.90E-1	2.23E-1	3.50E-1	2.68E-1	⋯	⋯	⋯	⋯	⋯	⋯	⋯	⋯
4600	⋯	⋯	2.82E-1	2.20E-1	3.81E-1	2.97E-1	2.17E-6	1.69E-6	⋯	⋯	⋯	⋯	⋯	⋯
4680	⋯	⋯	2.54E-1	2.01E-1	4.04E-1	3.19E-1	2.69E-6	2.13E-6	⋯	⋯	⋯	⋯	⋯	⋯
4760	⋯	⋯	2.05E-1	1.64E-1	3.94E-1	3.15E-1	⋯	⋯	⋯	⋯	⋯	⋯	⋯	⋯
4840	⋯	⋯	1.54E-1	1.24E-1	4.23E-1	3.42E-1	2.81E-3	2.27E-3	⋯	⋯	⋯	⋯	⋯	⋯
4920	⋯	⋯	1.02E-1	8.32E-2	3.92E-1	3.20E-1	1.40E-1	1.14E-1	⋯	⋯	⋯	⋯	⋯	⋯
5000	⋯	⋯	7.37E-2	6.05E-2	4.24E-1	3.48E-1	3.39E-1	2.79E-1	⋯	⋯	⋯	⋯	⋯	⋯
5080	⋯	⋯	4.37E-2	3.61E-2	4.29E-1	3.54E-1	4.05E-1	3.34E-1	⋯	⋯	⋯	⋯	⋯	⋯
5160	⋯	⋯	2.36E-2	1.96E-2	4.11E-1	3.41E-1	4.22E-1	3.50E-1	⋯	⋯	⋯	⋯	⋯	⋯
5240	⋯	⋯	1.07E-2	8.91E-3	4.15E-1	3.46E-1	4.21E-1	3.51E-1	⋯	⋯	⋯	⋯	⋯	⋯
5320	⋯	⋯	4.89E-3	4.10E-3	4.22E-1	3.54E-1	4.19E-1	3.51E-1	⋯	⋯	3.45E-6	2.89E-6	⋯	⋯
5400	⋯	⋯	3.07E-3	2.58E-3	2.56E-1	2.16E-1	3.76E-1	3.16E-1	8.77E-6	7.38E-6	1.30E-3	1.10E-3	⋯	⋯
5480	⋯	⋯	2.39E-3	2.01E-3	8.07E-2	6.81E-2	2.39E-1	2.01E-1	1.80E-4	1.52E-4	4.47E-2	3.77E-2	⋯	⋯
5560	⋯	⋯	2.02E-3	1.71E-3	5.58E-4	4.72E-4	1.18E-1	9.97E-2	4.87E-3	4.12E-3	1.81E-1	1.53E-1	⋯	⋯
5640	⋯	⋯	9.26E-4	7.85E-4	2.29E-6	1.95E-6	4.83E-2	4.10E-2	7.44E-2	6.31E-2	3.22E-1	2.73E-1	⋯	⋯
5720	⋯	⋯	2.58E-4	2.19E-4	⋯	⋯	2.19E-2	1.87E-2	2.44E-1	2.08E-1	4.11E-1	3.49E-1	⋯	⋯
5800	⋯	⋯	5.51E-5	4.70E-5	⋯	⋯	1.44E-2	1.23E-2	3.47E-1	2.96E-1	4.43E-1	3.77E-1	⋯	⋯
5880	⋯	⋯	1.06E-5	9.14E-6	⋯	⋯	1.08E-2	9.26E-3	3.82E-1	3.28E-1	4.49E-1	3.86E-1	⋯	⋯
5960	⋯	⋯	6.82E-6	5.90E-6	⋯	⋯	9.88E-3	8.55E-3	3.84E-1	3.33E-1	4.55E-1	3.94E-1	⋯	⋯

Table 10.6—Continued

λ (Å)	Bessel U 0.0	1.2	Bessel B 0.0	1.2	Sloan g' 0.0	1.2	Bessel V 0.0	1.2	R(λ) Bessel R 0.0	1.2	Sloan r' 0.0	1.2	Bessel I 0.0	1.2
6040	⋯	⋯	9.17E-6	8.02E-6	⋯	⋯	9.34E-3	8.17E-3	3.72E-1	3.26E-1	4.49E-1	3.93E-1	⋯	⋯
6120	⋯	⋯	6.01E-6	5.32E-6	⋯	⋯	5.25E-3	4.65E-3	3.71E-1	3.28E-1	4.47E-1	3.95E-1	⋯	⋯
6200	⋯	⋯	4.21E-6	3.76E-6	⋯	⋯	3.48E-3	3.11E-3	3.56E-1	3.18E-1	4.62E-1	4.12E-1	⋯	⋯
6280	⋯	⋯	2.60E-6	2.33E-6	⋯	⋯	2.54E-3	2.28E-3	3.33E-1	2.99E-1	4.35E-1	3.91E-1	⋯	⋯
6360	⋯	⋯	1.19E-6	1.07E-6	⋯	⋯	1.13E-3	1.02E-3	3.24E-1	2.92E-1	4.21E-1	3.80E-1	⋯	⋯
6440	⋯	⋯	⋯	⋯	⋯	⋯	6.82E-4	6.19E-4	3.01E-1	2.73E-1	4.34E-1	3.93E-1	⋯	⋯
6520	⋯	⋯	⋯	⋯	⋯	⋯	4.85E-4	4.42E-4	2.65E-1	2.41E-1	3.92E-1	3.57E-1	⋯	⋯
6600	⋯	⋯	⋯	⋯	⋯	⋯	1.99E-4	1.81E-4	2.48E-1	2.26E-1	3.73E-1	3.40E-1	⋯	⋯
6680	⋯	⋯	⋯	⋯	⋯	⋯	5.11E-5	4.68E-5	2.45E-1	2.24E-1	4.13E-1	3.78E-1	⋯	⋯
6760	⋯	⋯	⋯	⋯	⋯	⋯	1.30E-5	1.19E-5	2.34E-1	2.14E-1	4.40E-1	4.04E-1	⋯	⋯
6840	1.59E-6	1.42E-6	1.57E-6	1.40E-6	⋯	⋯	4.36E-6	3.88E-6	2.11E-1	1.88E-1	4.05E-1	3.60E-1	⋯	⋯
6920	9.02E-6	7.86E-6	1.31E-6	1.14E-6	⋯	⋯	1.06E-6	⋯	1.90E-1	1.65E-1	3.67E-1	3.20E-1	1.90E-6	1.65E-6
7000	3.06E-5	2.82E-5	1.04E-6	⋯	⋯	⋯	⋯	⋯	1.67E-1	1.54E-1	3.49E-1	3.22E-1	1.09E-3	1.01E-3
7080	6.20E-5	5.73E-5	⋯	⋯	⋯	⋯	⋯	⋯	1.42E-1	1.31E-1	2.15E-1	1.99E-1	1.70E-2	1.57E-2
7150	7.43E-5	6.55E-5	⋯	⋯	⋯	⋯	⋯	⋯	1.20E-1	1.06E-1	9.86E-3	8.68E-3	7.36E-2	6.48E-2
7240	5.59E-5	4.91E-5	⋯	⋯	⋯	⋯	⋯	⋯	1.02E-1	9.00E-2	5.15E-4	4.52E-4	1.61E-1	1.41E-1
7320	3.44E-5	3.11E-5	⋯	⋯	⋯	⋯	⋯	⋯	9.02E-2	8.17E-2	5.80E-5	5.25E-5	2.48E-1	2.25E-1
7400	1.88E-5	1.74E-5	⋯	⋯	⋯	⋯	⋯	⋯	7.83E-2	7.28E-2	1.10E-5	1.02E-5	3.12E-1	2.90E-1
7480	1.09E-5	1.01E-5	⋯	⋯	⋯	⋯	⋯	⋯	6.63E-2	6.17E-2	2.76E-6	2.56E-6	3.47E-1	3.23E-1
7560	5.55E-6	5.16E-6	⋯	⋯	⋯	⋯	⋯	⋯	5.45E-2	5.07E-2	1.17E-6	1.09E-6	3.58E-1	3.33E-1
7640	2.66E-6	2.47E-6	⋯	⋯	⋯	⋯	⋯	⋯	4.44E-2	2.46E-2	⋯	⋯	3.58E-1	1.98E-1
7720	1.17E-6	1.09E-6	⋯	⋯	⋯	⋯	⋯	⋯	3.58E-2	3.35E-2	⋯	⋯	3.53E-1	3.30E-1
7800	⋯	⋯	⋯	⋯	⋯	⋯	⋯	⋯	2.86E-2	2.68E-2	⋯	⋯	3.46E-1	3.24E-1
7850	⋯	⋯	⋯	⋯	⋯	⋯	⋯	⋯	2.28E-2	2.13E-2	⋯	⋯	3.39E-1	3.18E-1
7960	⋯	⋯	⋯	⋯	⋯	⋯	⋯	⋯	1.79E-2	1.67E-2	⋯	⋯	3.31E-1	3.11E-1
8040	⋯	⋯	⋯	⋯	⋯	⋯	⋯	⋯	1.39E-2	1.31E-2	⋯	⋯	3.23E-1	3.02E-1
8120	⋯	⋯	⋯	⋯	⋯	⋯	⋯	⋯	1.07E-2	9.80E-3	⋯	⋯	3.12E-1	2.84E-1
8200	⋯	⋯	⋯	⋯	⋯	⋯	⋯	⋯	8.16E-3	7.22E-3	⋯	⋯	3.00E-1	2.65E-1
8280	⋯	⋯	⋯	⋯	⋯	⋯	⋯	⋯	6.19E-3	5.48E-3	⋯	⋯	2.90E-1	2.57E-1
8360	⋯	⋯	⋯	⋯	⋯	⋯	⋯	⋯	4.75E-3	4.47E-3	1.14E-6	1.07E-6	2.85E-1	2.68E-1
8440	⋯	⋯	⋯	⋯	⋯	⋯	⋯	⋯	3.59E-3	3.38E-3	⋯	⋯	2.78E-1	2.62E-1
8520	2.13E-5	2.01E-5	3.41E-5	3.22E-5	⋯	⋯	2.60E-5	2.45E-5	2.94E-3	2.77E-3	1.34E-6	1.27E-6	2.71E-1	2.56E-1
8600	1.06E-5	9.97E-6	2.73E-5	2.57E-5	1.98E-6	1.86E-6	1.64E-5	1.55E-5	2.14E-3	2.02E-3	2.51E-5	2.37E-5	2.64E-1	2.49E-1
8760	1.18E-5	1.12E-5	2.18E-5	2.06E-5	1.93E-6	1.82E-6	1.90E-5	1.79E-5	1.23E-3	1.16E-3	2.01E-5	1.90E-5	2.45E-1	2.32E-1
8840	1.78E-5	1.68E-5	1.67E-5	1.58E-5	2.60E-6	2.46E-6	2.04E-5	1.92E-5	9.46E-4	8.93E-4	2.32E-5	2.19E-5	2.36E-1	2.23E-1
8920	7.39E-6	6.71E-6	1.45E-5	1.32E-5	1.89E-6	1.72E-6	1.60E-5	1.46E-5	7.12E-4	6.47E-4	2.19E-5	1.99E-5	2.29E-1	2.08E-1
9080	1.58E-5	1.35E-5	2.23E-5	1.91E-5	2.09E-6	1.79E-6	1.31E-5	1.12E-5	4.36E-4	3.73E-4	2.02E-5	1.73E-5	2.13E-1	1.82E-1

Table 10.6— Continued

λ Å	Bessel U 0.0	Bessel U 1.2	Bessel B 0.0	Bessel B 1.2	Sloan g' 0.0	Sloan g' 1.2	Bessel V 0.0	Bessel V 1.2	Bessel R 0.0	Bessel R 1.2	Sloan r' 0.0	Sloan r' 1.2	Bessel I 0.0	Bessel I 1.2
9160	1.25E-5	1.07E-5	9.32E-6	7.98E-6	1.29E-6	1.10E-6	1.34E-5	1.15E-5	3.45E-4	2.96E-4	1.60E-5	1.37E-5	2.05E-1	1.75E-1
9320	9.84E-6	6.19E-6	7.33E-6	4.61E-6	1.65E-6	1.04E-6	1.12E-5	7.06E-6	2.20E-4	1.39E-4	1.75E-5	1.10E-5	1.82E-1	1.15E-1
9400	1.63E-5	9.81E-6	1.44E-5	8.64E-6	1.80E-6	1.08E-6	8.83E-6	5.31E-6	1.89E-4	1.14E-4	1.71E-5	1.03E-5	1.71E-1	1.03E-1
9480	1.26E-5	7.56E-6	7.12E-6	4.28E-6	1.50E-6	:	2.07E-5	1.25E-5	1.19E-4	7.17E-5	1.37E-5	8.25E-6	1.59E-1	9.57E-2
9560	1.01E-5	6.47E-6	2.13E-5	1.37E-5	1.72E-6	1.10E-6	1.34E-5	8.60E-6	1.33E-4	8.57E-5	1.43E-5	9.16E-6	1.49E-1	9.55E-2
9640	1.17E-5	9.11E-6	1.37E-5	1.06E-5	1.60E-6	1.24E-6	1.24E-5	9.65E-6	9.09E-5	7.07E-5	1.41E-5	1.10E-5	1.34E-1	1.04E-1
9720	1.03E-5	8.99E-6	1.07E-5	9.37E-6	1.43E-6	1.25E-6	1.62E-5	1.42E-5	1.04E-4	9.10E-5	1.41E-5	1.23E-5	1.19E-1	1.05E-1
9800	3.12E-6	2.82E-6	6.75E-6	6.09E-6	1.07E-6	:	1.27E-5	1.15E-5	6.30E-5	5.68E-5	1.06E-5	9.56E-6	1.06E-1	9.61E-2
9880	1.48E-5	1.40E-5	8.37E-6	7.97E-6	1.34E-6	1.27E-6	1.29E-5	1.22E-5	7.13E-5	6.79E-5	1.35E-5	1.29E-5	9.38E-2	8.92E-2
10040	9.42E-6	8.97E-6	4.73E-6	4.50E-6	:	:	8.57E-6	8.15E-6	5.85E-5	5.57E-5	2.78E-5	2.65E-5	6.82E-2	6.49E-2
10120	3.66E-6	3.48E-6	1.20E-6	1.14E-6	:	:	8.53E-6	8.12E-6	3.80E-5	3.62E-5	6.37E-6	6.06E-6	5.77E-2	5.49E-2
10200	6.36E-6	6.06E-6	5.55E-6	5.28E-6	:	:	2.73E-6	2.60E-6	4.75E-5	4.52E-5	2.50E-4	2.38E-4	4.78E-2	4.55E-2
10360	3.14E-6	2.99E-6	4.95E-6	4.71E-6	:	:	5.93E-6	5.65E-6	2.28E-5	2.17E-5	1.76E-2	1.68E-2	3.16E-2	3.01E-2
10440	1.45E-6	1.38E-6	4.54E-6	4.32E-6	:	:	4.63E-6	4.41E-6	2.42E-5	2.30E-5	3.96E-3	3.77E-3	2.44E-2	2.32E-2
10600	2.46E-6	2.34E-6	2.68E-6	2.55E-6	:	:	:	:	8.08E-6	7.70E-6	9.43E-3	8.98E-3	1.32E-2	1.25E-2
10680	1.86E-6	1.77E-6	1.86E-6	1.77E-6	:	:	1.87E-6	1.79E-6	1.44E-5	1.37E-5	3.05E-2	2.90E-2	9.99E-3	9.51E-3
10760	1.20E-6	1.14E-6	1.58E-6	1.51E-6	:	:	2.09E-6	1.99E-6	1.24E-5	1.18E-5	9.46E-3	9.01E-3	7.25E-3	6.91E-3
10920	1.12E-6	1.07E-6	1.47E-6	1.40E-6	:	:	1.22E-6	1.16E-6	4.10E-6	3.90E-6	2.87E-3	2.74E-3	3.87E-3	3.68E-3
11000	:	:	:	:	:	:	1.15E-6	1.10E-6	3.91E-6	3.73E-6	2.90E-3	2.77E-3	2.80E-3	2.67E-3
11080	1.01E-6	:	1.10E-6	1.05E-6	:	:	:	:	4.94E-6	4.71E-6	3.72E-3	3.55E-3	1.92E-3	1.83E-3
11240	:	:	:	:	:	:	:	:	1.10E-6	1.05E-6	7.37E-3	7.03E-3	7.09E-4	6.77E-4
11320	:	:	:	:	:	:	:	:	1.87E-6	1.79E-6	4.27E-3	4.07E-3	3.89E-4	3.71E-4
11480	:	:	:	:	:	:	:	:	:	:	1.53E-3	1.46E-3	1.84E-4	1.75E-4
11560	:	:	:	:	:	:	:	:	:	:	1.35E-3	1.29E-3	1.38E-4	1.32E-4
11640	:	:	:	:	:	:	:	:	:	:	1.40E-3	1.34E-3	1.04E-4	9.92E-5
11720	:	:	:	:	:	:	:	:	1.49E-6	1.42E-6	1.66E-3	1.58E-3	7.52E-5	7.18E-5
11800	:	:	:	:	:	:	:	:	1.12E-6	1.07E-6	2.20E-3	2.10E-3	5.66E-5	5.41E-5
11880	:	:	:	:	:	:	:	:	:	:	3.04E-3	2.90E-3	4.15E-5	3.96E-5
11960	:	:	:	:	:	:	:	:	2.19E-6	2.10E-6	3.86E-3	3.69E-3	3.05E-5	2.91E-5

Note. — The calculated response curves are given for the 7 JCAM filters for two different airmasses. For clarity we suppress values of $R(\lambda) < 1 \times 10^{-6}$. The wavelength coverage of the response curves was chosen to coincide with the flux curves of the spectroscopic standards α Lyr (Vega) and BD17° 4708 that define the Sloan photometric system (see Fukugita et al. 1996). To reproduce the true total response on-sky imaging suggests that these numbers should be scaled downward by 3.170, 1.62, 1.53, 1.38, 1.75, 1.71, 3.54 in the filters U, B, V, g', R, r', and I, respectively. See text for an explanation.

CHAPTER 11

Epilogue and Future Steps

SECTION 11.1
On the Offset Distribution of GRBs

Since the offsets paper was completed (chapter 6), there have been a few new afterglow discoveries and *HST* observations of GRB hosts. The offsets and hosts for these are depicted in figure 11.1. The new bursts, two of which are the lowest-redshift GRBs measured, continue to show the close connection of GRBs to the light of host galaxies.

The claim in my offsets work (chapters 2 and 6) is that NS–NS and NS–BH mergers are inconsistent with the distribution of long-duration GRBs about their host galaxies. This statement holds only under the caveat that the calculated/expected radial models of such binaries are representative of the true distribution. The populations synthesis work in chapter 2 was the first attempt to produce such radial profiles in a variety of realistic galactic potentials. Since then a number of others (Fryer et al. 1999a; Bulik et al. 1999; Belczyński et al. 2000) have followed suit using different (and more sophisticated) population synthesis models for high-mass binary evolution. For the "standard" NS–NS production channel (Bhattacharya & van den Heuvel 1991; see fig. 1.3), all of our studies agree about offsets to within the uncertainties in input parameters[1], particularly the highly-uncertain supernova kick distribution (Fryer et al. 1999a). The three later studies also roughly agree on BH–NS offset distributions.

The comforting agreement between the different groups has recently been upset by the potential "discovery" of new channels of NS–NS production (Belczyński & Kalogera 2001; Belczynski et al. 2002b) that result in binaries which coalesce rapidly after ZAMS ($\lesssim 10^6$ yr). The existence of the dominant new channel relies on the untested assumption that low-mass helium stars can survive common-envelope evolution with a neutron star. If true, then the expected distribution of NS–NS mergers may be significantly more close to where massive stars are born (Belczynski et al. 2002a) and thus, by assumption, consistent with the observed offset distribution[2]. Moreover, given the close connection in time with star-formation, the redshift and the host-galaxy properties of such NS–NS systems would be indistinguishable from those properties expected from collapsar progenitors.

Even if such channels dominate the rate of NS–NS production, since neither NS is recycled and/or the binary merges quickly, it is unlikely that we will ever observe such a system in the Galaxy. We

---

[1] Much of the early NS–NS and NS–BH work was conducted primarily to predict the event rate for LIGO; these proceeded on both observational (e.g., Phinney 1991; Narayan et al. 1991; Kalogera & Lorimer 2000) and theoretical (population synthesis) (e.g., Tutukov & Yungelson 1993; Portegies Zwart & Spreeuw 1996; van den Heuvel & Lorimer 1996) grounds. Kalogera et al. (2001b), using the most up-to-date observations on known Galactic NS–NS binaries and pulsar statistics, has recently suggested that the birthrates of NS–NS binaries are still unknown by at least a factor of ~200. While this may be true, it is important to note that the uncertainty in the offset distribution should be much smaller. This is due to the somewhat counterintuitive fact, as I discovered in chapter 2, that different supernovae kick distributions only effect the birthrate of merging NS–NS, not the resulting distribution of systemic velocities.

[2] Note that BH–NS binaries are still excluded by the measured offset distribution since ~50% merger outside of 10 kpc.

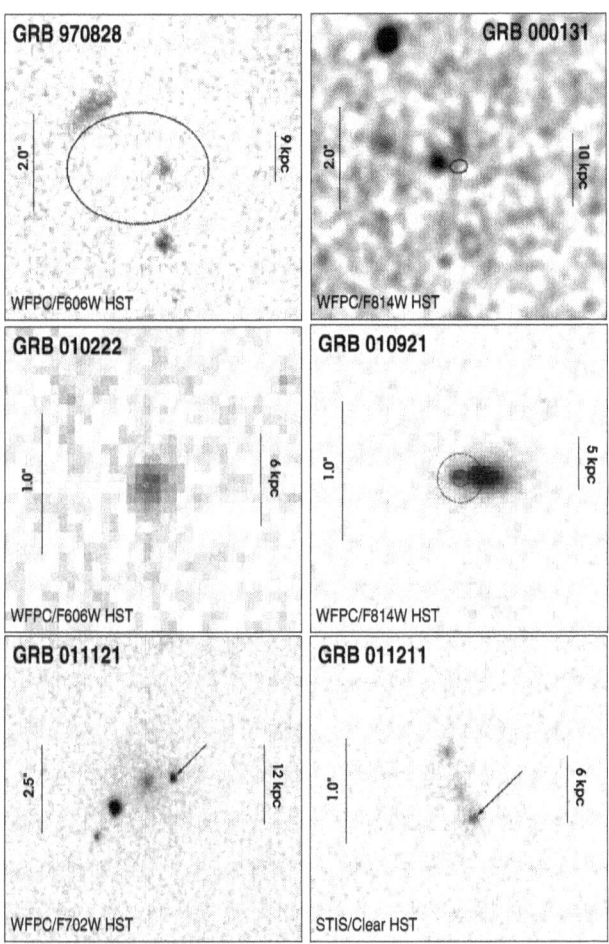

Figure 11.1 Update to chapter 6 of new offsets and *HST* images of host galaxies. (Top) 3 σ error contours are shown. (middle-left) The HST→HST astrometric tie places the OT very accurately ($\sigma_r = 6$ mas r.m.s.) on the host galaxy image of GRB 010222; displayed is the 10 σ error contour. (middle-right) The afterglow is still visible as a point source to the east of the apparent host galaxy; the error contours correspond to the 1 and 3 σ location predicted from an earlier Palomar image. (bottom) Arrows point to the optical transients.

are therefore left with the unsettling prospect of the existence of new theoretical channels for NS–NS births without obviously testable observations. A more detailed set of offset measurements might still be able to discriminate between progenitors[3] and more detailed hydrodynamical simulations should be employed of the common-envelope evolution phase to test the assumptions that lead to short-lived NS–NS binaries. The required number of new offsets ($\sim$50) to discriminate between this new NS–NS model and the collapsar model should be obtainable within the first $\sim$six months of the *Swift* mission even if the bursts occur in low-density environments (Perna & Belczynski 2002).

Perhaps more fruitful to distinguish between progenitors will be the use of high-quality early-time afterglow observations, afforded by rapid *Swift* localizations and instruments such as *JCAM*, to constrain the distribution of ambient density surrounding GRBs. In the rapid merger scenario, most bursts should still occur in low-density environments ($n \lesssim 1$ cm^{-3}; Perna & Belczynski 2002) whereas bursts from collapsars should occur with $n \gtrsim 1$ cm^{-3}. Already, afterglow modeling of some of the more extensive datasets have yielded densities far in excess of unity (Harrison et al. 2001; Panaitescu & Kumar 2001).

Aside from the continuing potential for offsets to directly discriminate between increasingly more sophisticated progenitors models, the continued measurement of offsets will be useful in new ways. Already the offsets work has shown that most long-duration GRBs probe the inner 10 kpc of their hosts. This, coupled with an expanded use of afterglow spectroscopic absorption line studies (Bloom et al. 2002), may shed new light on the enrichment history of moderate- to high-redshift galaxies in a manner complementary to quasar absorption-line studies (which probe the outer reaches of galaxies). Already we have seen that offsets plus spectroscopy can yield a lower-limit to the dynamical mass of GRB hosts (e.g., Castro et al. 2002).

SECTION 11.2
Re-examining the GRB–Supernova Connection

11.2.1 S-GRBs

In chapter 9, I suggested that there may be a sub-class of GRBs that are associated with local SNe. As evidenced by the exhaustive study undertaken by Norris et al. (1999) to find archival examples of S-GRBs in the *BATSE* catalog, the proposed sub-class has been taken seriously as a legitimate class of GRBs. In agreement with my estimates, those authors too found about 1–2% of the BATSE sample that met our S-GRB criteria. As expected by these estimates, no new examples of S-GRBs (as viewed purely from the γ-ray light curve considerations in chapter 9) have been found by *BeppoSAX* or the IPN since 1998.[4]

There have been some weak constraints on the S-GRB population placed by radio studies of Type Ib/Ic supernovae. None of the apparent optical hypernovae (i.e., SN 2002bl, SN 2002ap, SN 2002ao, SN 2002J, SN 2001bb, SN 1999as) were found to have prompt radio emission like 1998bw (though only 2002ap and 2002bl were promptly followed-up in the radio[5]; Berger, private communication). One of these (SN 2002ap) was also determined to have not produced a gamma-ray or X-ray burst (Hurley et al. 2002). These non-detections are consistent with our suggestion that only a fraction ($\sim$10%) of Type Ib/Ic supernovae could be truly the supernova component of an S-GRB.

The predictions of the S-GRB hypothesis could be tested by *Swift*. At current predicted rates (about

---

[3] Perna & Belczynski (2002), using these new channels, predicted the projected radial distribution of NS–NS and found all but $\sim$5–10 percent of NS–NS merge within 10 kpc of their host galaxy, depending on host mass. If the observed trend found in chapter 6, that all GRBs fall within 10 kpc of a galaxy, continues, then I calculate that we would need to observe between 26–72 (corresponding to a 10 and 5 percent extended population, respectively) to rule out the new NS–NS scenarios at the 99% confidence level.

[4] In retrospect, since supernovae remain bright (and detectable) for months in the optical, even delayed follow-up of old IPN localizations of GRBs might have uncovered a few supernova associations. Delayed searches were performed in the radio (Frail & Kulkarni 1995) but not systematically at optical wavelengths

[5] Four other non-hypernovae Type Ib/Ic supernovae were also followed-up in the radio with no detections (Berger, private communication).

Table 11.1. Summary of Proposed Cosmological GRBs with Associated Supernovae

Light Curve & Spectral Observations	Viable Alter. Models	Comments	Refs.
Good Candidates (definate bump detection)			
GRB 011121 Multi-color, multi-epoch observations of a bump before, during, and after peak. Shows evidence for spectral roll-over at $\sim 0.7 \mu$m.	?	Best case for a GRB–SN connection. Not an exact fit to SN 1998bw, but differences probably reflective of the diversity in core-collapased SNe.	1, 2
Plausible Candidates (probable bump detection)			
GRB 970228 V, I, R, K-band observations near peak. Broadband spectrum shows evidence for spectral roll-over at $\sim 0.9 \mu$m.	dust echo?	Good case for a GRB–SN connection, but spectral-roll over found from non-contemporaneous broadband obs. & extrapolations are highly uncertain.	3, 4, 5
GRB 980326 One R-band detection, few I-band upper-limits. Crude spectrum reveals red colors relative to the early afterglow.	dust echo	Best detection of extra emmission feature in R-band. Only redshift upper-limit. First case for GRB–SN connection.	6
GRB 990712 Multiple V-, R-band observations around SN peak, 2 HST epochs. No spectral info. Colors indicate red OT at $t = 48$ days.	dust echo, thermal dust	Faint SN is at comparable brightness to the host. Wrong SN colors. Fruchter et al. (2000c) finds no evidence for SN.	7, 8, 9
Marginal Candidates (marginal bump detection)			
GRB 000911 B, V, R, I, J-band obs near SN peak. Rebrightening claimed in R-, I-, J-band. Crude spectrum near SN peak.	dust echo	First claim of an IR brightening. Bright host yields 3σ detection of bump, but claim hinges on one J-band point. Host light contiminates spectrum.	10
Unlikely Candidates (improbable bump detection)			
GRB 991208 Few R, V-band observations around SN peak. No spectral info.	...	some evidence for bump in R-band but hinges on uncertain host flux.	11
GRB 970508 Few B, V, R, I-band observations around SN peak. No spectral info.	...	bump is a $\sim 2\sigma$ effect; only marginally seen in I-band.	12
GRB 000418 Few R-band observations around SN peak. No spectral info.	...	no apparent significance to claimed R-band bump. See ref. 14.	13, 14

Note. — Rank ordered list of possible GRBs with associated supernova signatures, from most likely to least likely. Ordering was based on estimated significance of bump detection (photometrically and spectroscopically) and number of viable alternative models to explain the detection(s).

References. — 1. Chapter 8; 2. Price et al. (2002); 3. Reichart (1999); 4. Galama et al. (2000); 5. Reichart (2001); 6. Bloom et al. (1999c) (chapter 7); 7. Sahu et al. (2000); 8. Fruchter et al. (2000c); 9. Björnsson et al. (2001); 10. Lazzati et al. (2001); 11. Castro-Tirado et al. (2001); 12. Sokolov (2001); 13. Dar & Rújula (2002); 14. Berger et al. (2001b)

two localizations per week), *Swift* should find about 1–3 S-GRBs per year down to the *BATSE* fluence limit ($S \sim 8 \times 10^{-7}$ erg cm^{-2}). Note that the *Swift* detection rate was based on extrapolated detection rates from *BATSE*. Since *Swift* is ~5 times more sensitive than *BATSE*, it is possible that the expected rate is a gross underestimate of the true rate—if the S-GRB hypothesis is correct, then not only will *Swift* detect bursts at a higher rate than presumed, but many of the faintest sources should be associated with supernovae out to a distance of ~225 h_{65}^{-1} Mpc, assuming that GRB 980425 was a standard candle. Even without decent localizations from *Swift*, a local population could be manifest in an upturn of the $\log N$–$\log S$ brightness distribution just fainter than the sensitivity of *BATSE*. Needless to say, if no S-GRBs are found by *Swift* after hundreds have been well-localized, the physical connection between 1998bw and GRB 980425 would be once again called in to question on statistical grounds.

Regardless of whether the S-GRB hypothesis is correct, there have been some tentative/weak suggestions of the existence of other local GRBs samples based upon gamma-ray properties:

- *Anisotropic Subsets*: The suggested angular anisotropy—indicative perhaps of a clustering in the super-galactic or Galactic plane—was found (e.g., Chen et al. 1998; Balazs et al. 1998; Mészáros et al. 2000) in some sub-classes that were created on purely phenomenological grounds (i.e., by choosing a locus of "intermediate" duration, soft-bursts). These findings are thus suspect because of the *ad hoc* creation of sub-sets of the data which yield the desired results.

- *Long-duration, long-lag bursts*: Norris (2002) has recently suggested a phenomenological sub-class of local GRBs motivated by the properties of GRB 980425 and other bursts with known (higher redshifts). The observed existence of the so-called luminosity–pulse lag relation, on which the possible local GRB sample is constructed, may even be motivated by a physical origin (Salmonson 2000).

- *X-ray Flashes (XRFs)*: Interestingly, XRFs detected by *BeppoSAX* are distributed such that $\langle V/V_{\max} \rangle = 0.50 \pm 0.09$ (Heise 1999), indicative of a distribution in Euclidean space; this would suggest a relatively local population.

11.2.2 Supernova bumps

The supernova interpretation for intermediate-time deviations of GRB afterglows from power-law behavior has become *en vogue* for new and historical GRBs but, apart from the recent discoveries of GRB 011121 (chapter 8), the observational connection has been largely disappointing. In table 11.1, I provide a summary of the GRBs with a proposed detection of a supernova component. These bursts are ranked in order of the security of the claimed detection as well as the number of plausible alternative models. The only comprehensive study which did not find a SN component (i.e., found an upper-limit on the peak brightness any associated SN) was from Price et al. (2002b), using GRB 010921.

The reasons for the lack of a convincing associations are many. First, most historical GRBs were poorly sampled at crucial time intervals: a GRB light curve must be well-sampled from about 1–3 months in order to conclusively see a SN signature; these time-scales were often not probed as extensively pre-GRB 980326. Second, there have been surprisingly few new GRBs since our discovery was published; only 14 bursts since October 1999 until March 2002 had optical or radio transients. This was due to a lower detection efficiency of *BeppoSAX* and the IPN as well as a delayed start to *HETE-II*. Third, SN signatures from GRBs which arise from redshifts higher than $z \sim 1$ are difficult to observe at optical wavelengths; only 5 of the 14 bursts remained as possible candidates. Last, of the low-redshift candidates, none of these bursts had a rapidly declining afterglow that would have improved the chance of detecting of an extra light curve component. These primarily observational impediments were ones which I anticipated in chapter 7.

Though many of the alternative models for the "red bump" in GRB 980326 were refuted in chapter 7, alternative explanations for the red bump were later put forth. Waxman & Draine (2000) suggested that the intermediate-time, red bumps could be produced by the thermal reemission of afterglow light from dust. While this cannot be ruled out in the case of GRB 980326, I suggested at the 5th Huntsville Conference in October 1999 that the spectral roll-over inferred in the red bump of GRB 970228 (Reichart 1999) excluded this model. Though the broadband spectrum of GRB 970228 was approximately thermal, the peak of the spectrum was at ~0.8 μm, while the peak of the thermal dust emission is expected at $\gtrsim 3\mu$m for the known redshift of GRB 970228. Reichart (2001) and Esin & Blandford (2000) later reiterated my point.

Esin & Blandford (2000) proposed that intermediate-time bumps could arise when dust around the GRB progenitor scatters the afterglow light into our light-of-sight. Assuming that the initial flash due to a reverse shock sublimates dust out to a radius R ($\approx$ 0.1–few pc; Waxman & Draine 2000), the timescale for the onset of emission due to dust scattering emission from the explosion site is $t \approx R/c \times (1+z) \times (1 - \cos\theta_{\rm typ}) \approx 10^6$ s (following eq. 6 of Esin & Blandford 2000). The typical scattering angle must be less than $\theta_{\rm typ} \lesssim 30$ deg for a high probability of scattering into the line-of-sight (see fig. 3 of Esin & Blandford 2000). Dust should preferentially scatter blue light more than red so that at low dust opacities, the intermediate-bump should appear more blue than the afterglow. At optical depths larger than $\tau_{0.3\mu m} \approx 3$, however, the absorption dominates and the bump appears more red than the afterglow (Esin & Blandford 2000). If the latter is true, as is claimed by Esin & Blandford (2000), then the intermediate-time emission of GRB 980326 and GRB 970228 can be explained as due to dust echoes and not an underlying supernova.

Reichart (2001) has called into question some of the conclusions of Esin & Blandford (2000). First, he claims the assumption of a thin scattering shell, left over by a wind-stratified medium after it has been sublimated, may be unwarranted because the radius of the scattering shell is required to be larger than the termination shock of the stellar wind from the progenitor. Second, he points out that the roll-over in the spectrum of GRB 970228 cannot be reproduced by dust scattering unless the roll-over pre-exists in the early afterglow itself (which it should not). Instead, spectra of dust-scattered light at a given time should be a monotonic function wavelength across the optical/infrared spectrum.

The discovery of a supernova-like component in GRB 011121 appears to disfavor the alternative models for red bumps. Though some may still be skeptical of the association, the timescale, spectra, and light curve of the GRB 011121 bump closely follows the simplistic expectations of a core-collapsed SN. Regardless, this clearly bodes well for future observational campaigns on low-redshift GRBs. What can be done to strengthen (or refute) the associations? The clearest, most unambiguous test for the GRB–SN connection will be obtaining a spectrum of an intermediate-time bump. At optical wavelengths, one expects to see redshifted broad metal-line absorption features from a Type Ib/Ic supernova (see chap. 7). Otherwise, in, for example, the dust echo origin, a smooth continuum is expected. This is a difficult observation from the ground. Ground-based spectroscopy of $R \approx 25$ mag bumps require a modest investment of observing time to see broad features from a point source ($\gtrsim$ 2 hr on a 10-m telescope in average seeing) but, due to atmospheric smearing of the bump and host light, will likely be hampered by the competing light of host galaxies. So unless the host is significantly fainter than the bump (as in the case of GRB 980326), even the detection of spectral features will lead to ambiguous conclusions. In principle, a careful subtraction of the light of the host (from a later-time spectrum after the bump has faded) could remove the contamination problem, but in practice, even if the observations were conducted using the same instrumental setup, it would be hard to exactly match the same observing conditions as the first epoch (e.g., seeing, airmass, slit location, etc.).

Space-based spectroscopy may be more productive and, just beginning in March 2002, feasible. In figure 11.2, I show a simulation of a supernova spectrum from the recently-installed *Advanced Camera for Surveys* (ACS) on the *HST*. In a few orbits ($\sim$ 4), broad SN features can be detected at better than 5σ for a $R = 26$ mag SN at $z = 1$. The angular resolution of the grism spectrum will be far superior than a ground-based spectrum, minimizing any host galaxy contamination. And, if any ambiguity still exists, a repeat observation at the same telescope roll-angle 6 months, or a 1 year later should allow a precise subtraction of the host contribution given the stability of the space-based instruments over time. It is not unreasonable to expect that such observations of bumps (and other faint point sources;

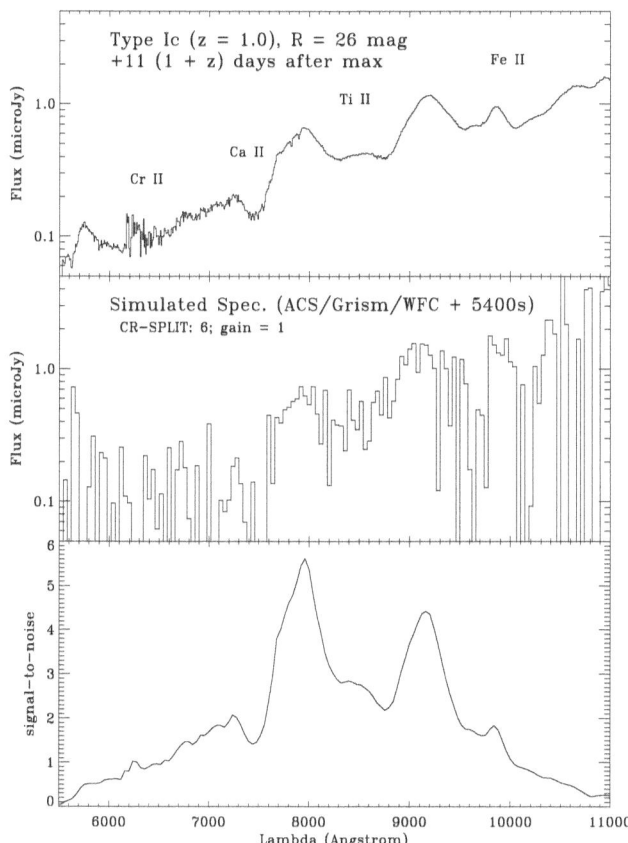

Figure 11.2 A future step toward resolving the progenitor question: space-based spectroscopy of intermediate-time emission components. Here, I show a simulated SNe spectrum at $R = 26$ mag and $z = 1.0$ as observed with the *Advanced Camera for Surveys* on *HST*. The template is SN 1994I (Type Ic) as reproduced in Millard et al. (1999). The observing parameters are given in the plot. Clearly the slope of the continuum is much steeper than $F_\nu \propto \nu^{-2}$ (as might be expected from competing hypotheses such as dust echos) and, more importantly, some broad spectral features are detectable even at such faint magnitude levels. The dispersion scale is 39 Å pixel^{-1}. The signal–to–noise on the bottom plot is given as the SN per resolution element, which in this case consists of 5 (spatial) × 2 (dispersion) = 10 pixels. The noise in the middle plot reflects the noise per pixel, summed over 5 pixels in the spatial direction (this encloses 88% of the total flux of a point source).

e.g., high-redshift SNe) will become one of the great uses of ACS, and *The Next Generation Space Telescope*. Another great hope for the next generation of sensitive space-based GRB instrumentation is that, thanks to an order of magnitude higher localization rate than currently, another apparently rare burst like GRB 980326 may be localized.

SECTION 11.3
Conclusions: What Makes Gamma-ray Bursts?

Table 11.2 depicts a critical (albeit cursory) assessment of the current state of viable progenitor scenarios in the face of existing observations. Dishearteningly, there is no obvious progenitor scenario which can naturally explain all of the observations of all of cosmological GRBs. Nevertheless, if we take the bold step and suggest that the phenomenological sub-classification of long-duration and short-duration bursts must actually represent a true distinction in progenitor models, then we are free to ask the question of what makes long-duration gamma-ray bursts.

The answer, which the reader should have seen growing clearer and clearer throughout the progress of this thesis, is that collapsars appear to be the only viable progenitor scenario which can explain most, if not all, of the data to date. The issue that gives us most pause is the disturbing lack of a detection of a wind-stratified medium in a GRB afterglow. Wind-stratified media are a natural expectation of the massive stellar progenitor scenarios. One suggestion is that there may be several different progenitors of long-duration GRBs (Chevalier & Li 1999; Livio & Waxman 2000) though it appears that almost all proposed wind-stratified bursts can be adequately modeled by a jetted burst in a constant-density medium (e.g., Frail et al. 2000c; Panaitescu & Kumar 2001). Instead, to explain the observations, one could invoke the case where the progenitor wind "turns-off" some $\lesssim 1000$ yrs before the GRB so that the dynamics of the afterglow are unaffected by the outwardly flowing wind. A wind-termination shock may also serve to homogenize the ambient medium surrounding a massive star progenitor (Wijers 2001). Scenarios where the wind is preferentially blown off along the equator and the burst is jetted along the polar axis also saves the collapsar.

Clearly, without an afterglow detection of a short burst, we have almost no observational evidence which constrains the progenitors of short-duration, hard-spectra GRBs (although see Hurley et al. 2002); however, the current theoretical picture, particularly related to the timescales for energy release, fosters the ample belief/hope that such bursts could be produced by merging remnants. The progenitors of short bursts will almost certainly be a subject of intensive study and discovery in the years to come.

11.3. CONCLUSIONS: WHAT MAKES GAMMA-RAY BURSTS?

Table 11.2. Summary Assessment of Progenitor Scenarios

Issue/ Requirement	Double Neutron Star Merger	Black-hole-Neutron Star Merger	Black-hole-Helium Star Merger	Black-hole-White Dwarf Merger	Collapsar	DRACO	AGN
Energetics (10^{51} of EM energy in a few secs)	Accretion-driven & spin energy driven possible.	Accretion-driven & spin energy driven possible.	Only spin energy driven possible (Narayan et al. 2001).	Only spin energy driven possible (Narayan et al. 2001).	Accretion-driven & spin energy driven possible.	Spin energy driven possible.	Not likely.
High Γ outflow?	Yes.	Yes.	Yes.	Yes.	Possibly. Baryon contamination a problem.	Yes.	Yes
Burst Durations	Cannot explain long-duration bursts.	Cannot explain long-duration bursts.	Difficult to explain short bursts?	Difficult to explain short bursts?	Only long-duration bursts.	Not well developed.	Unknown
Consistent with burst offsets?	No. Cannot explain observed distribution. See chapters 2 and 6.	Probably no.	Yes. Burst sites close to star-forming regions.	Probably no.	Yes. Burst sites inside star-forming regions.	Probably yes; burst sites near star-forming regions.	No. Should be at centers of hosts.
Consistent with absence of elliptical hosts?	Probably no.	Probably no.	Yes.	Yes.	Yes.	Yes.	No.
Explains transient iron lines?	No.	No.	Yes?	No.	Yes.	No.	No.
Explains intermediate-time afterglow bumps?	No.	No.	Yes, but contrived.	No.	Yes. A natural consequence of the model.	?	?
Explains apparent homogeneous ISM?	Yes.	Yes.	Yes, but requires kick outside of birthsite.	Yes.	Yes, but contrived ~1000 yr wind turn-off before burst.	?	?

Bibliography

Akerlof, C., et al. *Nature*, **398**, 400 (1999)

—. *ApJ*, **532**, L25 (2000)

Alcock, C., et al. *ApJ*, **521**, 602 (1999)

Anderson, J. & King, I. R. *PASP*, **111**, 1095 (1999)

Anderson, M. I. et al. *Science*, **283**, 2075 (1999)

Atteia, J.-L., et al. *ApJ*, **320**, L105 (1987)

Babul, A. & Ferguson, H. C. *ApJ*, **458**, 100 (1996)

Bagot, P., Portegies Zwart, S. F., & Yungelson, L. R. *A&A*, **332**, L57 (1998)

Balazs, L. G., Mészáros, A., & Horvath, I. *A&A*, **339**, 1 (1998)

Band, D. L. *ApJ*, **486**, 928 (1997)

Barbon, R., Cappellaro, E., & Turatto, M. *A&AS*, **81**, 421 (1989)

Barthelmy, S. D. In *Proc. SPIE X-Ray and Gamma-Ray Instrumentation for Astronomy XI*, (eds.) K. A. Flanagan & O. H. Siegmund, vol. 4140, pp. 50–63 (2000)

Bartunov, O. S., Tsvetkov, D. Y., & Filimonova, I. V. *PASP*, **106**, 1276 (1994)

Beckwith, S. *GRB 990123, Upcoming HST Service Observations*. GCN notice 245 (1999)

Begelman, M. C., Mészáros, P., & Rees, M. J. *MNRAS*, **265**, L13 (1993)

Belczyński, K., Bulik, T., & Zbijewski, W. *A&A*, **355**, 479 (2000)

Belczyński, K. & Kalogera, V. *ApJ*, **550**, L183 (2001)

Belczynski, K., Kalogera, V., & Bulik, T. *Study of Gamma Ray Burst Binary Progenitors*. Accepted to ApJ. astro-ph/0112122 (2002a)

Belczynski, K., Kalogera, V., & Bulik, V. *A Comprehensive Study of Binary Compact Objects as Gravitational Wave Sources: Evolutionary Channels, Rates, and Physical Properties*. Accepted to ApJ. astro-ph/0111452 (2002b)

Benetti, S., Pizzella, A., & Wheatley, P. J. *IAU Circ.* No. 6708 (1997a)

Benetti, S., Turatto, M., Perez, I., & Wisotzki, L. *IAU Circ.* No. 6554 (1997b)

Berger, E., Kulkarni, S. R., & Frail, D. A. *ApJ*, **560**, 652 (2001a)

Berger, E., et al. *ApJ*, **556**, 556 (2001b)

Bessell, M. S. & Brett, J. M. *PASP*, **100**, 1134 (1988)

Bethe, H. A. & Brown, G. E. ApJ, **506**, 780 (1998)

Bhattacharya, D. & van den Heuvel, E. P. J. Phys. Rep., **203**, 1 (1991)

Björnsson, G., Hjorth, J., Jakobsson, P., Christensen, L., & Holland, S. ApJ, **552**, L121 (2001)

Björnsson, G. & Lindfors, E. J. ApJ, **541**, L55 (2000)

Blaauw, A. Bulletin of the Astronomical Institute of the Netherlands, **15**, 265 (1961)

Blain, A. W. & Natarajan, P. MNRAS, **312**, L35 (2000)

Bloom, J. S. GRB 990510 Host Galaxy Discovery. GCN notice 756 (2000)

Bloom, J. S. GRB 020124: Optical observations. GCN notice 1225 (2002)

Bloom, J. S. & Berger, E. Fading of the optical source associated of XRF 011211. GCN notice 1193 (2001)

Bloom, J. S., Diercks, A., Kulkarni, S. R., Djorgovski, S. G., Scoville, N. Z., & Frayer, D. T. GRB 991208 Infrared Detection. GCN notice 480 (1999)

Bloom, J. S., Diercks, A., Kulkarni, S. R., Harrison, F. A., Behr, B. B., & Clemens, J. C. Optical observations of GRB 001018 with JCAM. GCN notice 915 (2001)

Bloom, J. S., Djorgovski, S. G., Gal, R. R., Kulkarni, S. R., & Kelly, A. GRB 980519 Optical Observations. GCN notice 87 (1998a)

Bloom, J. S., Djorgovski, S. G., & Kulkarni, S. R. ApJ, **554**, 678 (2001a)

Bloom, J. S., Djorgovski, S. G., Kulkarni, S. R., & Frail, D. A. ApJ, **507**, L25 (1998b)

Bloom, J. S., Fenimore, E. E., & in 't Zand, J. In Gamma Ray Bursts: 3rd Huntsville Symposium, (eds.) C. Kouveliotou, M. F. Briggs, & G. J. Fishman, vol. 384, p. 321 (Woodbury, New York: AIP, 1996)

Bloom, J. S., Frail, D. A., & Sari, R. AJ, **121**, 2879 (2001b)

Bloom, J. S., Kulkarni, S. R., & Djorgovski, S. G. AJ, **123**, 1111 (2002)

Bloom, J. S., Kulkarni, S. R., Djorgovski, S. G., & Frail, D. A. Optical Observations of GRB 970508. GCN notice 30 (1998c)

Bloom, J. S., Kulkarni, S. R., Djorgovski, S. G., Gal, R. R., Eichelberger, A., & Frail, D. A. Optical Observations of the Host Galaxy of GRB 980519. GCN notice 149 (1998d)

Bloom, J. S., Kulkarni, S. R., Harrison, F., Prince, T., Phinney, E. S., & Frail, D. A. ApJ, **506**, L105 (1998a)

Bloom, J. S., Kulkarni, S. R., et al. The Hosts of GRB 980703 and GRB 971214. GCN notice 702 (2000a)

Bloom, J. S., Sigurdsson, S., & Pols, O. R. MNRAS, **305**, 763 (1999a)

Bloom, J. S., Sigurdsson, S., Wijers, R. A. M. J., Almaini, O., Tanvir, N. R., & Johnson, R. A. MNRAS, **292**, L55 (1997)

Bloom, J. S., et al. ApJ, **508**, L21 (1998b)

—. ApJ, **518**, L1 (1999b)

—. Nature, **401**, 453 (1999c)

Bloom, J. S. et al. GRB 000418: Detection of the Host Galaxy. GCN notice 689 (2000b)

—. Absorption selected galaxies towards GRB 010222. In preparation (2002)

Boër, M., Atteia, J. L., Bringer, M., Gendre, B., Klotz, A., Malina, R., de Freitas Pacheco, J. A., & Pedersen, H. A&A, **378**, 76 (2001)

Boella, G., Butler, R. C., Perola, G. C., Piro, L., Scarsi, L., & Bleeker, J. A. M. *A&A*, **122**, 299 (1997)

Brainerd, J. J. In *Abstracts of the 19th Texas Symposium on Relativistic Astrophysics and Cosmology, held in Paris, France, Dec. 14–18, 1998*, (eds.) J. Paul, T. Montmerle, & E. Aubourg (Saclay: CEA, 1998)

Brandt, N. & Podsiadlowski, P. *MNRAS*, **274**, 461 (1995)

Brandt, W. N., Podsiadlowski, P., & Sigurdsson, S. *MNRAS*, **277**, L35 (1995)

Briggs, M. S., Pendleton, G. N., Kippen, R. M., Brainerd, J. J., Hurley, K., Connaughton, V., & Meegan, C. A. *Astrophys. J. Supp. Series*, **122**, 503 (1999)

Bruzual, A. G. & Charlot, S. *ApJ*, **405**, 538 (1993)

Buat, V., Deharveng, J. M., & Donas, J. *A&A*, **223**, 42 (1989)

Bulik, T., Belczyński, K., & Zbijewski, W. *MNRAS*, **309**, 629 (1999)

Burrows, D. N., et al. In *Proc. SPIE X-Ray and Gamma-Ray Instrumentation for Astronomy XI*, (eds.) K. A. Flanagan & O. H. Siegmund, vol. 4140, pp. 64–75 (2000)

Cardelli, J. A., Clayton, G. C., & Mathis, J. S. *ApJ*, **329**, L33 (1988)

Carter, B. *ApJ*, **391**, L67 (1992)

Castander, F. J. & Lamb, D. Q. *ApJ*, **523**, 602 (1999a)

—. *ApJ*, **523**, 593 (1999b)

Castro, S., Galama, T. J., Harrison, F. A., Holtzman, J. A., Bloom, J. S., Djorgovski, S. G., & Kulkarni, S. R. *Keck Spectroscopy and HST Imaging of GRB 000926: Probing a Host Galaxy at $z = 2.038$*. Submitted to ApJ; astro-ph/0110566 (2002)

Castro, S. M., Diercks, A., Djorgovski, S. G., Kulkarni, S. R., Galama, T. J., Bloom, J. S., Harrison, F. A., & Frail, D. A. *GRB 000301C: A Precise Redshift Determination*. GCN notice 605 (2000)

Castro-Tirado, A. J., et al. IAU Circ. No. 6848 (1998a)

Castro-Tirado, A. J., et al. *Science*, **279**, 1011 (1998b)

—. *A&A*, **370**, 398 (2001)

Celidonio, G., Coletta, A., Feroci, M., Piro, L., Soffitta, P., in 't Zand, J., Muller, J., & Palazzi, E. IAU Circ. No. 6851 (1998)

Chen, Y., Wu, M., & Song, L. M. *A&A*, **329**, 69 (1998)

Cheng, K. S. & Wang, J. *ApJ*, **521**, 502 (1999)

Chevalier, R. A. *ApJ*, **499**, 810 (1998)

Chevalier, R. A. & Li, Z. *ApJ*, **536**, 195 (2000)

Chevalier, R. A. & Li, Z.-Y. *ApJ*, **520**, L29 (1999)

Cline, T. L., et al. *A&AS*, **138**, 557 (1999)

Colgate, S. A. *Canadian Journal of Physics*, **46**, 476 (1968)

Cordes, J. M. & Chernoff, D. F. *ApJ*, **482**, 971 (1997)

Costa, E., et al. *Nature*, **387**, 783 (1997a)

Costa, E., et al. IAU Circ. No. 6576 (1997b)

Covino, S., et al. *A&A*, **348**, L1 (1999)

Crider, A., et al. *ApJ*, **479**, L39 (1997)

Dai, Z. G. & Lu, T. *A&A*, **333**, L87 (1998)

——. *ApJ*, **537**, 803 (2000)

Dar, A. & Rújula, A. D. *A cannonball model of gamma-ray bursts: superluminal signatures.* astro-ph/0008474 (2002)

Davis, L. *Specifications for the Aperature Photometry Package.* http://iraf.noao.edu/iraf/ftp/iraf/docs/apspec.ps.Z (1987)

de Bernardis, P. et al. *Nature*, **404**, 955 (2000)

de Vaucouleurs, G., de Vaucouleurs, A., Corwin, J. R., Buta, R. J., Paturel, G., & Fouque, P. *Third reference catalogue of bright galaxies* (New York: Springer-Verlag, 1991)

Dermer, C. D. In *Gamma Ray Bursts: 3rd Huntsville Symposium*, (eds.) C. Kouveliotou, M. F. Briggs, & G. J. Fishman, vol. 384, pp. 744–748 (Woodbury, New York: AIP, 1996)

Deutsch, E. W. *AJ*, **118**, 1882 (1999)

Dewey, R. J. & Cordes, J. M. *ApJ*, **321**, 780 (1987)

Diercks, A. et al. *GRB 991208 Host Galaxy Imaging.* GCN notice 764 (2000)

Djorgovski, S. G., Bloom, J. S., & Kulkarni, S. R. *The Redshift and the Host Galaxy of GRB 980613: A Gamma-Ray Burst From a Merger-Induced Starburst?* ApJ Lett., accepted; astro-ph/0008029 (2000)

Djorgovski, S. G., Frail, D. A., Kulkarni, S. R., Bloom, J. S., Odewahn, S. C., & Diercks, A. *ApJ*, **562**, 654 (2001)

Djorgovski, S. G., Kulkarni, S. R., Bloom, J. S., & Frail, D. *GRB 970228: Redshift and properties of the host galaxy.* GCN notice 289 (1999a)

Djorgovski, S. G., Kulkarni, S. R., Bloom, J. S., Goodrich, R., Frail, D. A., Piro, L., & Palazzi, E. *ApJ*, **508**, L17 (1998)

Djorgovski, S. G., Kulkarni, S. R., Bloom, J. S., S. C. Odewahn, R. R. G., & Frail, D. A. *GRB 990123: Possible Gravitationally Lensed Burst?* GCN notice 216 (1999b)

Djorgovski, S. G., Kulkarni, S. R., Cote, P., Blakeslee, J., Bloom, J. S., & Odewahn, S. C. *GRB 980326, optical observations.* GCN notice 57 (1998)

Djorgovski, S. G., Kulkarni, S. R., Gal, R. R., Odewahn, S. C., & Frail, D. A. IAU Circ. No. 6372 (1997a)

Djorgovski, S. G., et al. *Nature*, **387**, 876 (1997b)

——. *GRB 990123: Discovery of the Probable Host Galaxy.* GCN notice 256 (1999c)

——. *GRB 990705: Intriguing Positional Coincidences.* GCN notice 368 (1999d)

——. *The Cosmic Gamma-Ray Bursts.* To appear in: Proc. IX Marcel Grossmann Meeting, eds. V. Gurzadyan, R. Jantzen, and R. Ruffini, Singapore: World Scientific, in press (2001)

Djorgovski, S. G., et al. In *Gamma-Ray Bursts in the Afterglow Era, Proceedings of the International workshop held in Rome, CNR headquarters, 17–20 October, 2000*, (eds.) E. Costa, F. Frontera, & J. Hjorth, p. 218 (Berlin Heidelberg: Springer, 2001)

Dolphin, A. E. *PASP*, **112**, 1397 (2000)

Dutra, C. M., Bica, E., Clariá, J. J., Piatti, A. E., & Ahumada, A. V. *A&A*, **371**, 895 (2001)

Eichler, D., Livio, M., Piran, T., & Schramm, D. N. *Nature*, **340**, 126 (1989)

Epps, H. W. & Miller, J. S. *Proc. SPIE*, **3355**, 48 (1998)

Esin, A. A. & Blandford, R. *ApJ*, **534**, L151 (2000)

Fenimore, E. E. & Bloom, J. S. *ApJ*, **453**, 25 (1995)

Fenimore, E. E., Epstein, R. I., & Ho, C. *A&AS*, **97**, 59 (1993)

Fenimore, E. E., Ramirez-Ruiz, E., & Wu, B. *ApJ*, **518**, L73 (1999)

Fenimore, E. E., et al. *Nature*, **366**, 40 (1993)

Feroci, M., Piro, L., Frontera, F., Torroni, V., Smith, M., Heise, J., & in 't Zand, J. IAU Circ. No. 7095 (1999)

Filippenko, A. V. IAU Circ. No. 6783 (1997)

Filippenko, A. V. & Matheson, T. IAU Circ. No. 5740 (1993)

Fillipenko, A. V. *Ann. Rev. Astr. Ap.*, **35**, 309 (1997)

Finger, G., Biereichel, P., Mehrgan, H., Meyer, M., Moorwood, A. F., Nicolini, G., & Stegmeier, J. *Proc. SPIE*, **3354**, 87 (1998)

Fishman, G. J. & Meegan, C. A. *Ann. Rev. Astr. Ap.*, **33**, 415 (1995)

Folkes, S., et al. *MNRAS*, **308**, 459 (1999)

Frail, D. A. *GRB 991208: Radio Observations*. GCN notice 451 (1999)

Frail, D. A. & Kulkarni, S. R. *Astrophy. and Space Sci.*, **231**, 277 (1995)

Frail, D. A. & Kulkarni, S. R. *GRB 990123, a new radio source*. GCN notice 211 (1999)

Frail, D. A., Kulkarni, S. R., Nicastro, S. R., Feroci, M., & Taylor, G. B. *Nature*, **389**, 261 (1997)

Frail, D. A., Kulkarni, S. R., Shepherd, D. S., & Waxman, E. *ApJ*, **502**, L119 (1998)

Frail, D. A., Waxman, E., & Kulkarni, S. R. *ApJ*, **537**, 191 (2000a)

Frail, D. A., et al. *ApJ*, **525**, L81 (1999)

—. In *Gamma Ray Bursts: 5th Huntsville Symposium*, (ed.) G. J. F. R. Marc Kippen, Robert S. Mallozzi, vol. 526, p. 298 (Meville, New York: AIP, 2000b)

—. *ApJ*, **534**, 559 (2000c)

—. *ApJ*, **562**, L55 (2001)

Freedman, W. L., et al. *ApJ*, **427**, 628 (1994)

Frontera, F., Costa, E., dal Fiume, D., Feroci, M., Nicastro, L., Orlandini, M., Palazzi, E., & Zavattini, G. *A&A*, **122**, 357 (1997)

Fruchter, A., Bergeron, L., & Pian, E. IAU Circ. No. 6674 (1997)

Fruchter, A. & Hook, R. N. In *Applications of Digital Image Processing XX, Proc. SPIE, Vol. 3164*, (ed.) A. Tescher, pp. 120–125 (SPIE, 1997)

Fruchter, A., Hook, R. N., Busko, I. C., & Mutchier, M. In *The 1997 HST Calibration Workshop with a new generation of instruments*, p. 518 (Baltimore, MD : Space Telescope Science Institute, 1997)

Fruchter, A., Metzger, M., Petro, L., et al. *GRB 000301C: Further Late-Time HST/STIS observations*. GCN notice 701 (2000a)

Fruchter, A. & Pian, E. *GRB 970508 optical observations.* GCN notice 151 (1998)

Fruchter, A., Sahu, K., Gibbons, R., et al. *GRB 990712 HST Observations.* GCN notice 575 (2000b)

Fruchter, A., Thorsett, S., Pian, E., et al. *Late-time HST/STIS Observation of GRB 990123.* GCN notice 354 (1999)

Fruchter, A., Vreeswijk, P., Hook, R., et al. *GRB 990712: Late time HST/STIS Observations.* GCN notice 752 (2000c)

Fruchter, A., Vreeswijk, P., & Nugent, P. *GRB 980326: Late-time HST/STIS observations.* GCN notice 1029 (2001a)

Fruchter, A., Vreeswijk, P., Sokolov, V., & Castro-Tirado, A. *GRB 991208: HST Imaging of the Host Galaxy.* GCN notice 872 (2000d)

Fruchter, A., Vreeswijk, P., et al. *GRB 000301C: Late-time HST/STIS observation.* GCN notice 1063 (2001b)

Fruchter, A., et al. *ApJ*, **516**, 683 (1999a)

—. *ApJ*, **519**, L13 (1999b)

Fruchter, A. et al. *Late-time HST Observations of GRB 990510.* GCN notice 386 (1999)

—. *GRB 000301C.* GCN notice 627 (2000e)

Fryer, C., Burrows, A., & Benz, W. *ApJ*, **496**, 333 (1998)

Fryer, C. L. & Woosley, S. E. *ApJ*, **502**, L9 (1998)

Fryer, C. L., Woosley, S. E., & Hartmann, D. H. *ApJ*, **526**, 152 (1999a)

Fryer, C. L., Woosley, S. E., Herant, M., & Davies, M. B. *ApJ*, **520**, 650 (1999b)

Fukugita, M., Ichikawa, T., Gunn, J. E., Doi, M., Shimasaku, K., & Schneider, D. P. *AJ*, **111**, 1748 (1996)

Fukugita, M., Shimasaku, K., & Ichikawa, T. *PASP*, **107**, 945 (1995)

Fynbo, J. P. U. et al. *GRB 000301C: Optical Candidate.* GCN notice 570 (2000)

Fynbo, J. P. U., et al. *ApJ*, **542**, L89 (2000)

Galama, T. *Gamma-Ray Burst Afterglows.* Ph.D. thesis, University of Amsterdam (1999)

Galama, T., et al. *Nature*, **387**, 479 (1997)

Galama, T. J., Vreeswijk, P. M., Pian, E., Frontera, F., Doublier, V., & Gonzalez, J.-F. IAU Circ. No. 6895 (1998)

Galama, T. J., et al. *ApJ*, **497**, L13 (1998a)

—. *Nature*, **395**, 670 (1998b)

Galama, T. J. et al. *GRB 990510 Optical Observations.* GCN 313 (1999)

Galama, T. J., et al. *A&A*, **138**, 465 (1999)

—. *ApJ*, **536**, 185 (2000)

Garcia, M. R., et al. *ApJ*, **500**, L105 (1998)

Garnavich, P. M., Holland, S. T., Jha, S., Kirshner, R. P., Bersier, D., & Stanek, K. Z. *GRB 011121 possible supernova association.* GCN notice 1273 (2002)

Garnavich, P. M., Loeb, A., & Stanek, K. Z. *ApJ*, **544**, L11 (2000)

Gehrels, N. A. In *Proc. SPIE X-Ray and Gamma-Ray Instrumentation for Astronomy XI*, (eds.) K. A. Flanagan & O. H. Siegmund, vol. 4140, pp. 42–49 (2000)

Germany, L. M., Reiss, D. J., Sadler, E. M., Schmidt, B. P., & Stubbs, C. W. *ApJ*, **533**, 320 (2000)

Ghisellini, G. & Lazzati, D. *MNRAS*, **309**, L7 (1999)

González, R. A., Fruchter, A. S., & Dirsch, B. *ApJ*, **515**, 69 (1999)

Goodman, J. *ApJ*, **308**, 47 (1986)

Greiner, J. *Astrophy. and Space Sci.*, **231**, 263 (1995)

Groot, P. J., et al. *ApJ*, **502**, L123 (1998a)

—. *ApJ*, **493**, L27 (1998b)

Gubler, J. & Tytler, D. *PASP*, **110**, 738 (1998)

Höflich, P., Wheeler, J. C., & Wang, L. *ApJ*, **521**, 179 (1999)

Hakkila, J., Meegan, C. A., Pendleton, G. N., Briggs, M. S., Horack, J. M., Hartmann, D. H., & Connaughton, V. In *Gamma Ray Bursts: 4th Huntsville Symposium*, (eds.) C. A. Meegan, R. Preece, & T. Koshut, vol. 428, pp. 236–240 (Woodbury, New York: AIP, 1998)

Halpern, J. P., Kemp, J., Piran, T., & Bershady, M. A. *ApJ*, **517**, L105 (1999)

Hansen, B. M. S. *ApJ*, **512**, L117 (1999)

Hansen, B. M. S. & Phinney, E. S. *MNRAS*, **291**, 569 (1997)

Harding, A. K. In *Supernovae and Gamma-Ray Bursts: the Greatest Explosions since the Big Bang*, (ed.) K. S. Mario Livio, Nino Panagia, vol. 13, p. 121 (Cambridge, UK: Cambridge University Press, 2001)

Harrison, F. A., et al. *ApJ*, **523**, L121 (1999)

—. *ApJ*, **559**, 123 (2001)

Hartman, J. W., Bhattacharya, D., Wijers, R., & Verbunt, F. *A&A*, **322**, 477 (1997)

Heise, J. *Jan van Paradijs Memorial Conference; Amsterdam, June 2001*. Conference presentation (1999)

Heise, J., in 't Zand, J. J. M., Kulkarni, S., & Costa, E. *The X-ray Flash 011030 ('GRB 011030')*. GCN notice 1138 (2001)

Hernquist, L. *ApJ*, **356**, 359 (1990)

Hjorth, J., Andersen, M. I., Pedersen, H., Jaunsen, A. O., Costa, E., & Palazzi, E. *GRB 980613 Optical Observations*. GCN notice 109 (1998)

Hjorth, J., Andersen, M. I., Pedersen, H., Zapatero-Osorio, M., Perz, E., & Castro-Tirado, A. J. *GRB 990123 NOT Spectrum Update*. GCN notice 249 (1999)

Hjorth, J., Holland, S., Courbin, F., Dar, A., Olsen, L. F., & Scodeggio, M. *ApJ*, **534**, L147 (2000)

Hjorth, J. et al. *GRB 981226, HST/STIS observations of the host galaxy*. GCN notice 749 (2000)

Hogg, D. W., Cohen, J. G., Blandford, R., & Pahre, M. A. *ApJ*, **504**, 622 (1998)

Hogg, D. W. & Fruchter, A. *ApJ*, **520**, 54 (1999)

Hogg, D. W., Pahre, M. A., McCarthy, J. K., Cohen, J. G., Blandford, R., Smail, I., & Soifer, B. T. *MNRAS*, **288**, 404 (1997)

Holland, S., Andersen, M., & Hjorth, J. *GRB 990705, HST/STIS observations of the host galaxy*. GCN notice 793 (2000a)

Holland, S., Fynbo, J., Thomsen, B., et al. *GRB 980425, HST/STIS observations of the host galaxy*. GCN notice 704 (2000b)

—. *GRB 980519, HST/STIS observations of the host galaxy*. GCN notice 698 (2000c)

Holland, S. & Hjorth, J. *A&A*, **344**, L67 (1999)

Holland, S., Thomsen, B., Bjornsson, G., et al. *GRB 990506, HST/STIS Observations of the Host Galaxy*. GCN notice 731 (2000d)

Holland, S., et al. *GRB 990308: HST/STIS observations of the host galaxy*. GCN notice 726 (2000e)

—. *HST/STIS observations of the chaotic environment of GRB 980613*. GCN notice 777 (2000f)

Holland, S. et al. *HST/STIS observations of the host galaxy of GRB 980329*. GCN notice 778 (2000g)

Holtzman, J. A., Burrows, C. J., Casertano, S., Hester, J. J., Trauger, J. T., Watson, A. M., & Worthey, G. *PASP*, **107**, 1065 (1995)

Hughes, D. H., et al. *Nature*, **394**, 241 (1998)

Hurley, K. *A Gamma-Ray Burst Bibliography, 1973–2001*. To appear in the proceedings of the conference on Gamma-Ray Burst and Afterglow Astronomy 2001; astro-ph/0201350 (2002)

Hurley, K., Hartmann, D., Kouveliotou, C., Fishman, G., Laros, J., Cline, T., & Boer, M. *ApJ*, **479**, L113 (1997)

Hurley, K. et al. *IPN triangulation of GRB 000630*. GCN notice 736 (2000a)

—. *IPN triangulation of GRB 000911*. GCN notice 791 (2000b)

—. *IPN triangulation of GRB 000926*. GCN notice 801 (2000c)

Hurley, K., et al. *ApJ*, **567**, 447 (2002)

Hurley, K. et al. *IPN Upper Limits to a GRB Associated With SN2002ap*. GCN notice 1252 (2002)

Infante, L., Garnavich, P. M., Stanek, K. Z., & Wyrzykowski, L. *GRB 011121: possible redshift, continued decay*. GCN notice 1152 (2001)

Iwamoto, K., et al. *Nature*, **395**, 672 (1998)

Jager, R., et al. *A&AS*, **125**, 557 (1997)

Janka, H.-T., Eberl, T., Ruffert, M., & Fryer, C. L. *ApJ*, **527**, L39 (1999)

Jaunsen, A. O., Hjorth, J., Andersen, M. I., et al. *GRB 980519 Optical Observations*. GCN notice 78 (1998)

Johnson, H. M. & MacLeod, J. M. *PASP*, **75**, 123 (1963)

Johnston, H. M. & Kulkarni, S. R. *ApJ*, **368**, 504 (1991)

Kalogera, V. & Lorimer, D. R. *ApJ*, **530**, 890 (2000)

Kalogera, V., Narayan, R., Spergel, D. N., & Taylor, J. H. *ApJ*, **556**, 340 (2001a)

—. *ApJ*, **556**, 340 (2001b)

Katz, J. I. *ApJ*, **432**, L107 (1994)

Kells, W., Dressler, A., Sivaramakrishnan, A., Carr, D., Koch, E., Epps, H., Hilyard, D., & Pardeilhan, G. *PASP*, **110**, 1487 (1998)

Kennicut, R. C. *Ann. Rev. Astr. Ap.*, **36**, 131 (1998)

Kennicutt, R. C. *ApJ*, **344**, 685 (1989)

Khokhlov, A. M., Höflich, P. A., Oran, E. S., Wheeler, J. C., Wang, L., & Chtchelkanova, A. Y. *ApJ*, **524**, L107 (1999)

Kim, A., Goobar, A., & Perlmutter, S. *PASP*, **108**, 190 (1996)

Kimble, R. A., et al. *ApJ*, **492**, L83 (1998)

Kippen, R. M., et al. *ApJ*, **506**, L27 (1998)

Kirshner, R. P., et al. *ApJ*, **415**, 589 (1993)

Klebesadel, R. W., Strong, I. B., & Olson, R. A. *ApJ*, **182**, L85 (1973)

Kluźniak, W. & Ruderman, M. *ApJ*, **505**, L113 (1998)

Kobayashi, S. & Sari, R. *ApJ*, **551**, 934 (2001)

Kouveliotou, C. et al. In *Gamma Ray Bursts: 3rd Huntsville Symposium*, (eds.) C. Kouveliotou, M. F. Briggs, & G. J. Fishman, vol. 384, pp. 42–46 (Woodbury, New York: AIP, 1996)

Kulkarni, S. R., Bloom, J. S., Frail, D. A., Ekers, R., Wieringa, M., Wark, R., Higdon, J. L., & Monard, B. IAU Circ. No. 6903 (1998a)

Kulkarni, S. R., Djorgovski, S. G., Clemens, J. C., Gal, R. R., Odewahn, S. C., & Frail, D. A. IAU Circ. No. 6732 (1997)

Kulkarni, S. R., et al. *Nature*, **393**, 35 (1998b)

Kulkarni, S. R., et al. *Nature*, **395**, 663 (1998)

Kulkarni, S. R., et al. *Nature*, **398**, 389 (1999a)

—. *ApJ*, **522**, L97 (1999b)

—. *Proc. SPIE*, **4005**, 9 (2000)

Kumar, P. *ApJ*, **523**, L113 (1999)

Lamb, D. Q. *PASP*, **107**, 1152 (1995)

Landolt, A. U. *AJ*, **104**, 372 (1992)

Larkin, J., Ghez, A., Kulkarni, S., Djorgovski, S., Frail, D., & Taylor, G. *GRB 980329 Keck K-band observations*. GCN notice 51 (1998)

Lazzati, D., Covino, S., & Ghisellini, G. In *Gamma Ray Bursts: 5th Huntsville Symposium*, (ed.) G. J. F. R. Marc Kippen, Robert S. Mallozzi, vol. 526, p. 318 (Meville, New York: AIP, 2000)

—. *MNRAS*, **330**, 583 (2002)

Lazzati, D., et al. *A&A*, **378**, 996 (2001)

Le Fèvre, O., et al. *MNRAS*, **311**, 565 (2000)

Levine, A. M., Bradt, H., Cui, W., Jernigan, J. G., Morgan, E. H., Remillard, R., Shirey, R. E., & Smith, D. A. *ApJ*, **469**, L33 (1996)

Li, L.-X. & Paczyński, B. *ApJ*, **507**, L59 (1998)

Lide, D. R. *CRC Handbook of Chemistry and Physics* (1994)

Lilly, S. J., Tresse, L., Hammer, F., Crampton, D., & Le Fevre, O. *ApJ*, **455**, 108 (1995)

Lindegren, L. *A&A*, **89**, 41 (1980)

Lipunov, V. M. *Relativistic Binary Merging Rates*. astro-ph/9711270 (1997)

Lipunov, V. M., Postnov, K. A., & Prokhorov, M. E. *MNRAS*, **288**, 245 (1997)

Lipunov, V. M., Postnov, K. A., Prokhorov, M. E., Panchenko, I. E., & Jorgensen, H. E. *ApJ*, **454**, 593 (1995)

Livio, M. & Waxman, E. *ApJ*, **538**, 187 (2000)

Lloyd, N. M. & Petrosian, V. *ApJ*, **511**, 550 (1999)

Loeb, A. *Institute for Theoretical Physics, University of California, Santa Barbara*. Conference presentation (1999)

Lund, N. *A&AS*, **231**, 217 (1995)

Lyne, A. G. & Lorimer, D. R. *Nature*, **369**, 127 (1994)

Mészáros, A., Bagoly, Z., Horváth, I., Balázs, L. G., & Vavrek, R. *ApJ*, **539**, 98 (2000)

Mészáros, P., Laguna, P., & Rees, M. J. *ApJ*, **415**, 181 (1993)

Mészáros, P. & Rees, M. J. *ApJ*, **556**, L37 (2001)

MacFadyen, A. I. & Woosley, S. E. *ApJ*, **524**, 262 (1999)

MacFadyen, A. I., Woosley, S. E., & Heger, A. *ApJ*, **550**, 410 (2001)

Madau, P. IAU Circ. No. 186 (1997)

Madau, P., Pozzetti, L., & Dickinson, M. *ApJ*, **498**, 106 (1998)

Mallozzi, R. S., Paciesas, W. S., Pendleton, G. N., Briggs, M. S., Preece, R. D., Meegan, C. A., & Fishman, G. J. *ApJ*, **454**, 597 (1995)

Malumuth, E. M. & Bowers, C. W. In *The 1997 HST Calibration Workshop with a new generation of instruments*, (eds.) S. Casertano, R. Jedrzejewski, C. D. Keyes, & M. Stevens, p. 144 (Baltimore, MD: Space Telescope Science Institute, 1997)

Mao, S. & Mo, H. J. *A&A*, **339**, L1 (1998)

Masetti, N., et al. *A&A*, **354**, 473 (2000)

Massey, P., Strobel, K., Barnes, J. V., & Anderson, E. *ApJ*, **328**, 315 (1988)

Matthews, K. & Soifer, B. T. In *Infrared Astronomy with Arrays, the Next Generation*, (ed.) I. McLean, p. 239 (Dordrecht: Kluwer, 1994)

Mazzali, P. A., et al. *ApJ*, **572**, L61 (2002)

McKenzie, E. H. & Schaefer, B. E. *PASP*, **111**, 964 (1999)

McLean, I. S., et al. *SPIE*, **3354**, 566 (1998)

Meegan, C. A., Fishman, G. J., Wilson, R. B., Horack, J. M., Brock, M. N., Paciesas, W. S., Pendleton, G. N., & Kouveliotou, C. *Nature*, **355**, 143 (1992)

Meegan, C. A. et al. In *Gamma Ray Bursts: 4th Huntsville Symposium*, (eds.) C. A. Meegan, R. Preece, & T. Koshut, vol. 428, pp. 3–9 (Woodbury, New York: AIP, 1998)

Mészáros, P. & Rees, M. J. *MNRAS*, **257**, 29P (1992)

Mészáros, P. & Rees, M. J. *ApJ*, **418**, L59 (1993)

Mészáros, P. & Rees, M. J. *ApJ*, **405**, 278 (1993)

Mészáros, P. & Rees, M. J. *ApJ*, **476**, 232 (1997a)

—. *ApJ*, **482**, L29 (1997b)

Mészáros, P., Rees, M. J., & Papathanassiou, H. *ApJ*, **432**, 181 (1994)

Mészáros, P., Rees, M. J., & Wijers, R. A. M. J. *ApJ*, **499**, 301 (1998)

Metzger, M., Fruchter, A., Masetti, N., et al. *GRB 000418: HST/STIS observations*. GCN notice 733 (2000)

Metzger, M. R., Cohen, J. G., Chaffee, F. H., & Blandford, R. D. IAU Circ. No. 6676 (1997a)

Metzger, M. R., Djorgovski, S. G., Kulkarni, S. R., Steidel, C. C., Adelberger, K. L., Frail, D. A., Costa, E., & Fronterra, F. *Nature*, **387**, 879 (1997b)

Metzger, M. R., Kulkarni, S. R., Djorgovski, S. G., Gal, R., Steidel, C. C., & Frail, D. A. IAU Circ. No. 6588 (1997c)

Millard, J., et al. *ApJ*, **527**, 746 (1999)

Mirabal, N., Halpern, J. P., Wagner, R. M., et al. *GRB 000418, Optical Observation*. GCN notice 650 (2000)

Mochkovitch, R., Hernanz, M., Isern, J., & Martin, X. *Nature*, **361**, 236 (1993)

Monet, D. G. In *American Astronomical Society Meeting*, vol. 193, p. 12003 (1998)

Nakamura, T. *Progress of Theoretical Physics*, **100**, 921 (1998)

Narayan, R. & Ostriker, J. P. *ApJ*, **352**, 222 (1990)

Narayan, R., Paczyński, B., & Piran, T. *ApJ*, **395**, L83 (1992)

Narayan, R., Piran, T., & Kumar, P. *ApJ*, **557**, 949 (2001)

Narayan, R., Piran, T., & Shemi, A. *ApJ*, **379**, L17 (1991)

Natarajan, P., et al. *New Astronomy*, **2**, 471 (1997)

Nemiroff, R. J. *PASP*, **107**, 1131 (1995)

Nicklas, H., Seifert, W., Boehnhardt, H., Kiesewetter-Koebinger, S., & Rupprecht, G. *Proc. SPIE*, **2871**, 1222 (1997)

Norris, J. P. *Implications of Lag-Luminosity relationship for unified GRB paradigms*. astro-ph/0201503 (2002)

Norris, J. P., Bonnell, J. T., & Watanabe, K. *ApJ*, **518**, 901 (1999)

Odewahn, S. C., Bloom, J. S., & Kulkarni, S. R. IAU Circ. No. 7094 (1999)

Odewahn, S. C., Windhorst, R. A., Driver, S. P., & Keel, W. C. *ApJ*, **472**, L13 (1996a)

—. *ApJ*, **472**, L13 (1996b)

Odewahn, S. C., et al. *ApJ*, **509**, L5 (1998)

Oke, J. B. & Gunn, J. E. *ApJ*, **266**, 713 (1983)

Oke, J. B., et al. *PASP*, **107**, 375 (1995)

Olsen, K., Brown, M., Schommer, R., & Stubbs, C. *GRB 011121*. GCN Notice 1157 (2001)

Östlin, G., Amram, P., Bergvall, N., Masegosa, J., Boulesteix, J., & Márquez, I. *A&A*, **374**, 800 (2001)

Paciesas, W. S., et al. *ApJS*, **122**, 465 (1999)

Paczyński, B. *ApJ*, **308**, L43 (1986)

Paczyński, B. *ApJ*, **335**, 525 (1988)

—. *ApJ*, **348**, 485 (1990)

Paczyński, B. *PASP*, **107**, 1167 (1995)

—. *ApJ*, **494**, L45 (1998)

Paczyński, B. & Rhoads, J. E. *ApJ*, **418**, L5 (1993)

Panaitescu, A. & Kumar, P. *ApJ*, **543**, 66 (2000)

—. *ApJ*, **560**, L49 (2001)

Panaitescu, A., Mészáros, P., & Rees, M. J. *ApJ*, **503**, 314 (1998)

Park, H. S., et al. In *Gamma Ray Bursts: 5th Huntsville Symposium*, (ed.) G. J. F. R. Marc Kippen, Robert S. Mallozzi, vol. 526, p. 736 (Meville, New York: AIP, 2000)

Pedersen, H. et al. *ApJ*, **496**, 311 (1998)

Pendleton, G. N. et al. *ApJ*, **489**, 175 (1997)

Perlmutter, S., et al. *ApJ*, **483**, 565 (1997)

Perna, R. & Belczynski, K. *ApJ*, **570**, 252 (2002)

Peters, P. C. *Physical Review*, **136**, 1224 (1964)

Phinney, E. S. *ApJ*, **380**, L17 (1991)

Pian, E. *Optical Observations of Afterglows*. To appear in "Supernovae and Gamma-Ray Bursters," edited by K. W. Weiler, Springer-Verlag Press (2001)

Pian, E., et al. *ApJ*, **492**, L103 (1998)

Pian, E., et al. *A&A*, **138**, 463 (1999)

Piran, T. *Phys. Rep.*, **314**, 575 (1999)

Piran, T., Kumar, P., Panaitescu, A., & Piro, L. *ApJ*, **560**, L167 (2001)

Piro, L. et al. *GRB 990123, BeppoSAX-NFI X-ray afterglow detection*. GCN notice 203 (1999a)

—. *GRB 990123, BeppoSAX WFC detection and NFI planned follow-up*. GCN notice 199 (1999b)

Piro, L., et al. *ApJ*, **514**, L73 (1999)

—. *Science*, **290**, 955 (2000)

Piro, L. et al. *BeppoSAX/GRB 011121*. GCN notice 1147 (2001a)

—. *GRB 010222: the brightest GRB observed by BeppoSAX*. GCN notice 959 (2001b)

Pols, O. R. & Marinus, M. *A&A*, **288**, 475 (1994)

Porciani, C. & Madau, P. *ApJ*, **548**, 522 (2001)

Portegies Zwart, S. F. & Spreeuw, H. N. *A&A*, **312**, 670 (1996)

Portegies Zwart, S. F. & Verbunt, F. *A&A*, **309**, 179 (1996)

Portegies Zwart, S. F. & Yungelson, L. R. *A&A*, **332**, 173 (1998)

Price, P. A., Schmidt, B. P., & Kulkarni, S. R. *GRB 010921: HST Observations*. GCN notice 1259 (2002a)

Price, P. A. et al. *GRB 010921*. In preparation (2002b)

Price, P. A., et al. *ApJ*, **572**, L51 (2002)

Price, P. A. et al. *ApJ*, **572**, L51 (2002)

Ramirez-Ruiz, E., Dray, L. M., Madau, P., & Tout, C. A. *MNRAS*, **327**, 829 (2001)

Ramirez-Ruiz, E. & Fenimore, E. E. *ApJ*, **539**, 712 (2000)

Rana, N. C. & Wilkinson, D. A. *MNRAS*, **218**, 497 (1986)

Readhead, A. C. S. *ApJ*, **426**, 51 (1994)

Reaves, G. *PASP*, **65**, 242 (1953)

Rees, M. J. *A&AS*, **138**, 491 (1999)

Rees, M. J. & Mészáros, P. *ApJ*, **430**, L93 (1994)

Rees, M. J. & Mészáros, P. *ApJ*, **496**, L1 (1998)

Reichart, D. E. *ApJ*, **521**, L111 (1999)

Reichart, D. E. *Dust Echoes*. Submitted to ApJ Letters. (2001)

Reichart, D. E. *ApJ*, **554**, 643 (2001)

Reshetnikov, V. P. *Astronomy Letters*, **26**, 61 (2000)

Rhoads, J. E. *apj*, **525**, 737 (1999)

Ricker, G. R. & HETE Science Team. *American Astronomical Society Meeting*, **198** (2001)

Roche, N., Shanks, T., Metcalfe, N., & Fong, R. *MNRAS*, **280**, 397 (1996)

Rol, E., et al. *ApJ*, **544**, 707 (2000)

Rupen, M. P., Sramek, R. A., van Dyk, S. D., Weiler, K. W., Panagia, N., Richmond, M. W., Filippenko, A. V., & Treffers, R. R. IAU Circ. No. 5963 (1994)

Sadler, E. M., Stathakis, R. A., Boyle, B. J., & Ekers, R. D. IAU Circ. No. 6901 (1998)

Sahu, K. C., et al. *Nature*, **387**, 476 (1997)

Sahu, K. C., et al. *ApJ*, **540**, 74 (2000)

Salmonson, J. D. *ApJ*, **544**, L115 (2000)

—. *ApJ*, **546**, L29 (2001)

Sarl, R. *ApJ*, **524**, L43 (1999)

Sari, R. & Piran, T. *ApJ*, **517**, L109 (1999a)

—. *A&A*, **138**, 537 (1999b)

Sari, R., Piran, T., & Halpern, J. P. *ApJ*, **519**, L17 (1999)

Sari, R., Piran, T., & Narayan, R. *ApJ*, **497**, L17 (1998)

Schaefer, B. E., et al. *ApJ*, **313**, 226 (1987)

—. *ApJ*, **524**, L103 (1999)

Schlegel, D. J., Finkbeiner, D. P., & Davis, M. *ApJ*, **500**, 525 (1998)

Schmidt, B. P., et al. *ApJ*, **507**, 46 (1998)

Schmidt, M. *ApJ*, **523**, L117 (1999)

—. *ApJ*, **559**, L79 (2001)

Sigurdsson, S. & Rees, M. J. *MNRAS*, **284**, 318 (1997)

Simard, L., et al. *ApJ*, **519**, 563 (1999)

Smette, A., et al. *ApJ*, **556**, 70 (2001)

Smith, J. A., et al. *AJ*, **123**, 2121 (2002)

Soderberg, A. M. & Ramirez-Ruiz, E. *MNRAS*, **330**, L24 (2002)

Sokolov, V. V. *Bull. Spec. Astrophys. Obs.*, **51**, 38 (2001)

Sokolov, V. V., Kopylov, A. I., Zharikov, S. V., Feroci, M., Nicastro, L., & Palazzi, E. *A&A*, **334**, 117 (1998)

Sokolov, V. V., et al. *A&A*, **372**, 438 (2001)

Stanek, K. Z., Garnavich, P. M., Kaluzny, J., Pych, W., & Thompson, I. *ApJ*, **522**, L39 (1999)

Stanek, K. Z., Garnavich, P. M., & Wyrzykowski, L. *GRB 01121: fading bahavior*. GCN notice 1151 (2001)

Stanek, K. Z. & Wyrzykowski, L. *GRB 011121: fainter still*. GCN Notice 1160 (2001)

Stasinska, G. *A&AS*, **83**, 501 (1990)

Steidel, C. C. & Sargent, W. L. W. *ApJS*, **80**, 1 (1992)

Stetson, P. B. *PASP*, **99**, 191 (1987)

Stone, R. C. *AJ*, **97**, 1227 (1989)

Stornelli, M., Celidonio, G., Muller, J., Zand, J. I., Amati, L., Feroci, M., & Gandolfi, G. *GRB 000210: BeppoSAX GRBM and WFC Detection*. GCN notice 540 (2000)

Strong, I. B., Klebesadel, R. W., & Olson, R. A. *ApJ*, **188**, L1 (1974)

Sutantyo, W. *Astrophysics and Space Science*, **54**, 479 (1978)

Tammann, G. A., Loeffler, W., & Schroeder, A. *Astrophys. J. Supp. Series*, **92**, 487 (1994)

Taylor, G. B., Beasley, A. J., Frail, D. A., Kulkarni, S. R., & Reynolds, J. E. *A&A*, **138**, 445 (1999)

Taylor, G. B., Bloom, J. S., Frail, D. A., Kulkarni, S. R., Djorgovski, S. G., & Jacoby, B. A. *ApJ*, **537**, L17 (2000)

Taylor, G. B. et al. *ApJ*, **502**, L115 (1998)

Taylor, J. H. *Rev. Mod. Phys.*, **66** (3), 711 (1994)

Taylor, J. H. & Dewey, R. J. *ApJ*, **332**, 770 (1988)

Taylor, J. H. & Weisberg, J. M. *ApJ*, **345**, 434 (1989)

Thompson, R. I., Storrie-Lombardi, L. J., Weymann, R. J., Rieke, M. J., Schneider, G., Stobie, E., & Lytle, D. *AJ*, **117**, 17 (1999)

Tinney, C. et al. IAU Circ. No. 6896 (1998)

Tonry, J. L., Hu, E. M., Cowie, L. L., & McMahon, R. G. IAU Circ. No. 6620 (1997)

Totani, T. ApJ, **486**, L71 (1997)

Tutukov, A. V. & Yungelson, L. R. MNRAS, **260**, 675 (1993)

—. MNRAS, **268**, 871 (1994)

Uglesich, R., Mirabal, N., Halpern, J., Kassin, S., & Novati, S. GRB 991216, Optical Afterglow. GCN notice 472 (1999)

Usov, V. V. Nature, **357**, 472 (1992)

Valdes, F., Jannuzi, B., & Rhoads, J. GRB 980326, optical observations. GCN notice 56 (1998)

van den Bergh, S. & Tammann, G. A. Ann. Rev. Astr. Ap., **29**, 363 (1991)

van den Heuvel, E. P. J. & Lorimer, D. R. MNRAS, **283**, L37 (1996)

van Dyk, S. D. AJ, **103**, 1788 (1992)

van Dyk, S. D., Sramek, R. A., Weiler, K. W., & Panagia, N. ApJ, **409**, 162 (1993)

van Paradijs, J., Kouveliotou, C., & Wijers, R. A. M. J. Ann. Rev. Astr. Ap., **38**, 379 (2000)

van Paradijs, J., et al. Nature, **386**, 686 (1997)

Vanderspek, R., Krimm, H. A., & Ricker, G. R. In Gamma-Ray Bursts: Proceedings of the 2nd Workshop, (ed.) G. J. Fishman, vol. 307, p. 438 (New York: AIP, 1994)

Verbunt, F., Wijers, R. A. M. J., & Burm, H. M. G. A&A, **234**, 195 (1990)

Vietri, M. ApJ, **478**, L9 (1997)

Vietri, M., Perola, C., Piro, L., & Stella, L. MNRAS, **308**, L29 (1999)

Vietri, M. & Stella, L. ApJ, **507**, L45 (1998)

Vrba, F. J., Hartmann, D. H., & Jennings, M. C. ApJ, **446**, 115 (1995)

Vreeswijk, P. M., Fruchter, A., Ferguson, H., Kouveliotou, C., et al. GRB 991216, HST/STIS observations. GCN notice 751 (2000)

Vreeswijk, P. M., Galama, T. J., Rol, E., et al. GRB 990510 optical observation. GCN notice 310 (1999)

Walborn, N. R., Prevot, M. L., Prevot, L., Wamsteker, W., Gonzalez, R., Gilmozzi, R., & Fitzpatrick, E. L. A&A, **219**, 229 (1989)

Wang, L., Howell, D. A., & Wheeler, J. C. IAU Circ. No. 6802 (1998)

Wang, L. & Wheeler, J. C. ApJ, **504**, L87 (1998)

Wang, X. & Loeb, A. ApJ, **535**, 788 (2000)

Wax, N. (ed.). Selected Papers on Noise And Stochastic Processes (Dover Publications, Inc., 1954). See pg. 238

Waxman, E. ApJ, **489**, L33 (1997a)

—. ApJ, **485**, L5 (1997b)

Waxman, E. & Draine, B. T. ApJ, **537**, 796 (2000)

Waxman, E., Kulkarni, S. R., & Frail, D. A. ApJ, **497**, 288 (1998)

Waxman, E. & Loeb, A. *ApJ*, **515**, 721 (1999)

Weiler, K. W. & Sramek, R. A. *Ann. Rev. Astr. Ap.*, **26**, 295 (1998)

Weth, C., Mészáros, P., Kallman, T., & Rees, M. J. *ApJ*, **534**, 581 (2000)

Wettig, T. & Brown, G. E. *New Astronomy*, **1**, 17 (1996)

Wheeler, J. C. In *Supernovae and Gamma-Ray Bursts: the Greatest Explosions since the Big Bang*, (ed.) K. S. Mario Livio, Nino Panagia, vol. 13, p. 356 (Cambridge, UK: Cambridge University Press, 2001)

Wheeler, J. C., Harkness, R. P., Clocchiatti, A., Benetti, S., Brotherton, M. S., Depoy, D. L., & Elias, J. *ApJ*, **436**, L135 (1994)

White, G. L. & Malin, D. F. *Nature*, **327**, 36 (1987)

Wieringa, M. et al. IAU Circ. No. 6896 (1998)

Wieringa, M. H., Kulkarni, S. R., & Frail, D. A. *A&A*, **138**, 467 (1999)

Wijers, R. A. M. J. In *Gamma-Ray Bursts in the Afterglow Era, Proceedings of the International workshop held in Rome, CNR headquarters, 17–20 October, 2000*, (eds.) E. Costa, F. Frontera, & J. Hjorth, p. 306 (Berlin Heidelberg: Springer, 2001)

Wijers, R. A. M. J., Bloom, J. S., Bagla, J., & Natarajan, P. *MNRAS*, **294**, L17 (1998)

Wijers, R. A. M. J., Rees, M. J., & Mészáros, P. *MNRAS*, **288**, L51 (1997)

Wijers, R. A. M. J., van Paradijs, J., & van den Heuvel, E. P. J. *A&A*, **261**, 145 (1992)

Wijers, R. A. M. J., et al. *ApJ*, **523**, L33 (1999)

Williams, R. E., et al. *AJ*, **112**, 1335 (1996)

Wolszczan, A. *Nature*, **350**, 688 (1991)

Woods, E. & Loeb, A. *ApJ*, **508**, 760 (1998)

Woosley, S. E. *ApJ*, **405**, 273 (1993)

Woosley, S. E., Eastman, R. G., & Schmidt, B. P. *ApJ*, **516**, 788 (1999)

Wyrzykowski, L., Stanek, K. Z., & Garnavich, P. M. *GRB 011121: possible optical counterpart*. GCN notice 1150 (2001)

Yoshida, A., Namiki, M., Otani, C., Kawai, N., Murakami, T., Ueda, Y., Shibata, R., & Uno, S. *A&A*, **138**, 433 (1999)

Zharikov, S. V., Sokolov, V. V., & Baryshev, Y. V. *A&A*, **337**, 356 (1998)

www.ingramcontent.com/pod-product-compliance
Lightning Source LLC
Chambersburg PA
CBHW030937180526
45163CB00002B/598